Regional Conference in Civil Engineering and Sustainable Development Goals in Higher Education Institutions (RCCE SDG 2020)

IOP Conference Series: Materials Science and Engineering
Volume 1153

Johor, Malaysia
23 - 25 January 2021

ISBN: 978-1-7138-3423-6
ISSN: 1757-8981

Printed from e-media with permission by:

Curran Associates, Inc.
57 Morehouse Lane
Red Hook, NY 12571

CURRAN ASSOCIATES INC. proceedings.com

Some format issues inherent in the e-media version may also appear in this print version.

Institute of Physics
Dirac House, Temple Back
Bristol BS1 6BE UK

Phone: 44 1 17 929 7481
Fax: 44 1 17 920 0979

techtracking@iop.org

Additional copies of this publication are available from:

Curran Associates, Inc.
57 Morehouse Lane
Red Hook, NY 12571 USA
Phone: 845-758-0400
Fax: 845-758-2633
Email: curran@proceedings.com
Web: www.proceedings.com

TABLE OF CONTENTS

IOP Conf. Series: Materials Science and Engineering 1153 (2021) 011001 doi:10.1088/1757-899X/1153/1/011001

Editorial Material

1. Preface/Introduction

With great pleasure, we would like to present to you the scholarly selected articles presented in the Regional Conference in Civil Engineering and Sustainable Development Goals in Higher Education Institutions 2020 (RCCE 2020), held between 23rd and 24th January 2021 conducted virtually via Whova Virtual Conference Platform in Universiti Teknologi Malaysia, Johor, Malaysia. This conference was sponsored by ASEAN Universiti Network/Southeast Asia Engineering Education Development Network (AUN/SEED-NET) under the Japan International Cooperation Agency (JICA) Technical Cooperation Project.

RCCE 2020 strategized to combine the transformation of Sustainable Development Goals (SDGs) implementation in Higher Education Institutions (HEIs) based on the requirements of United Nation for the year 2030 as well as to support the specific agenda in "The World in 2050" (TWI 2050). This aspiration was the global research and fact-based knowledge in transforming the indicated SDGs in a better way. Therefore, all the keynotes/ invited/ guest speakers explained their roles and contribution to the society, with SDGs related to R&D, innovation, technology and knowledge-transfer. Most of the issues and challenges were connected to the environmental application thereby strengthening the interaction between HEIs, Industry, Government and Community.

As the world is still bound to the COVID-19 pandemic, travelling was restricted. For this, different parallel sessions were conducted virtually focusing on the theme "Localizing Sustainable Development Goals (SDGs) in Engineering with Look East Policy Partnership". Participants around this region did not only share similar issues or problems faced but also forged strong partnership among the ASEAN countries to come out with viable solutions fitting this region.

In line with the international publication requirements, manuscripts from sessions related to engineering field including structural, environmental, material science, green technology, geotechnical, transportation, water resources and wastewater were rigorously reviewed. We hope that the collection of accepted manuscript will provide insights to many academicians, researchers and specialist of particular field.

2. Conference or sponsor logo(s)

Conference Logo

Sponsor Logo

3. List of committees

International Scientific Committee	
Prof. Dr. Iwanami Mitsuyasu	Tokyo Institute of Technology, Japan
Prof. Dr. Asakura Yasuo	Tokyo Institute of Technology, Japan
Prof. Dr. Kanae Shinjiro	Tokyo Institute of Technology, Japan
Prof. Dr. Yokota Hiroshi	Hokkaido University, Japan
Prof. Dr. Teerapong Senjuntichai	Chulalongkorn University, Thailand
Assoc. Prof. Dr. Tirawat Boonyatee	Chulalongkorn University, Thailand
Assoc. Prof. Dr. Zainah Ibrahim	University of Malaya
Asst. Prof. Dr. Watanachai Smittakorn	Chulalongkorn University, Thailand

Technical Committee	
Prof. Ir. Dr. Mohd Fadhil Md Din	Universiti Teknologi Malaysia
AP Ir. Ts. Dr. Ahmad 'Athif Mohd Faudzi	Universiti Teknologi Malaysia
AP Dr. Mohd Hafiz Dzarfan Othman	Universiti Teknologi Malaysia
AP Dr. Noorminshah binti A.iahad	Universiti Teknologi Malaysia
AP Dr. Norhazilan Bin Md. Noor	Universiti Teknologi Malaysia
Dr. Afikah Rahim	Universiti Teknologi Malaysia
Dr. Afnizanfaizal Bin Abdullah	Universiti Teknologi Malaysia
Dr. Dianah Mazlan	Universiti Teknologi Malaysia
Dr. Eeydzah Aminudin	Universiti Teknologi Malaysia
Dr. Farhana Aziz	Universiti Teknologi Malaysia
Dr. Fadzlin Md Sairan	Universiti Teknologi Malaysia
Dr. Khairun Nissa Mat Said	Universiti Teknologi Malaysia
Dr. Kogila Vani Annammala	Universiti Teknologi Malaysia
Dr. Norasikin Saman	Universiti Teknologi Malaysia
Dr. Nor Eliza Alias	Universiti Teknologi Malaysia
Dr. ONG Pei Ying	Universiti Teknologi Malaysia
Dr. Shazwin Mat Taib	Universiti Teknologi Malaysia
Dr. Tan Joo Siang	Universiti Teknologi Malaysia
Dr. Yong Ee Ling	Universiti Teknologi Malaysia

Non-technical Committee

Mdm. Nurfarhain Md Rusli	Universiti Teknologi Malaysia
Mdm. Tan Hooi Luang	Universiti Teknologi Malaysia
Ms. Leelavathi A/P Samikan Mazilamani	Universiti Teknologi Malaysia
Ms. Liang Yun Qi	Universiti Teknologi Malaysia
Ms. Nazleatul Najiha M. Nazif	Universiti Teknologi Malaysia
Ms. Nurul Syafiqa Mohd Yusof	Universiti Teknologi Malaysia
Ms. Nurulhuda binti Radznam	Universiti Teknologi Malaysia
Ms. Putri Ma'Inah Rosdin	Universiti Teknologi Malaysia
Ms. Rasnavi a/p Paramasivam	Universiti Teknologi Malaysia
Ms. Siti Nur Syamimi Mohd Na'im	Universiti Teknologi Malaysia
Mr. Fauzan Fahmi Mohamed Nora'eni	Universiti Teknologi Malaysia
Mr. Mohammad Amirul Fitry Mohd Bahar	Universiti Teknologi Malaysia
Mr. Muhammad Wafiy Adli Ramli	Universiti Teknologi Malaysia
Mr. Muhamad Hairulnizam Ishak	Universiti Teknologi Malaysia
Mr. Norhisyam Hanafi	Universiti Teknologi Malaysia
Mr. Wan Lokman Wan Bedurdin	Universiti Teknologi Malaysia

IOP Conf. Series: Materials Science and Engineering 1153 (2021) 011002 doi:10.1088/1757-899X/1153/1/011002

Peer review declaration

All papers published in this volume of *IOP Conference Series: Materials Science and Engineering* have been peer reviewed through processes administered by the Editors. Reviews were conducted by expert referees to the professional and scientific standards expected of a proceedings journal published by IOP Publishing.

- **Type of peer review: Single-blind**
  Peer review conducted was a single-blind review.

- **Conference submission management system:**
  Easychair

- **Number of submissions received:**
  85

- **Number of submissions sent for review:**
  43 (Engineering related manuscripts)

- **Number of submissions accepted:**
  20

- **Acceptance Rate (Number of Submissions Accepted / Number of Submissions Received X 100):**
  23.5%

- **Average number of reviews per paper:**
  1.3

- **Total number of reviewers involved:**
  31

- **Any additional info on review process:**
  Reviewers are to adhere to the requirements below:-

  1. All unpublished manuscripts are confidential documents.
  2. Equal consideration shall be given to all articles regardless of race, gender, religion, ethnicity, citizenship or political philosophy of the authors.
  3. Any terminology used shall not be offensive or might be perceived to be offensive to others.
  4. Statement of recommendation for each article shall be clear and comments shall be supplied in support of the recommendation.
  5. Only articles with clear support from reviewers will be accepted for publication.
  6. If plagiarism is detected in the article reviewed, please kindly notify the publication committee.

IOP Conf. Series: Materials Science and Engineering 1153 (2021) 011002 doi:10.1088/1757-899X/1153/1/011002

- **Contact person for queries:**
  Dr Yong Ee Ling (email: eeling@utm.my)
  Dr Nor Eliza Alias (email: noreliza@utm.my)

Evaluation of Ground Motion Definition on the Seismic Analysis of Reinforced Concrete Moment Frame Structure

T Ornthammarath and K M Myint

Department of Civil and Environmental Engineering, Faculty of Engineering, Mahidol University
Email: teraphan.orn@mahidol.edu

Abstract. The primary objective of the study is to analyze the effect of the horizontal definition in the ground motion selection, in which, how the target spectra of DBE and MCE resulted from the geometric mean method and NGA West-2 $Sa_{RotD100}$, the maximum direction method is different from each other. The analysis of the seismic performance of the building is the secondary aim of the research. The linear design methods of equivalent lateral force analysis, response spectrum and seismic response history analysis are performed based on the Myanmar seismic design code to figure out and compare the building responses such as story displacement, story drift, story base shear, and story overturning moment. In this study, the irregular shaped RC building structural responses are also explored with equivalent lateral force analysis (EQLF), response spectrum analysis (RSA) and linear response history analysis (LRHA). The recent ground motion definition response shows great potential in improving the performance-based design of considered structures.

1. Introduction

The static and dynamic structural analysis become primary concern in earthquake engineering to predict the structural behavior subjected to earthquake ground motions. The ground motion selection becomes one of the important issues in the seismic analysis to perform dynamic analysis. An earthquake is a natural disaster and can happen naturally without warning. Moreover, the structure tends to collapse and it may harm the life of people during earthquake excitations. Thus, the proper design for the structure is needed. The seismic provision codes are proposed to ensure the structure and occupant benefits in each country respectively. Since Myanmar is one of the earthquake-prone zones in Southeast Asia, the seismic performance analysis is recommended in this region. Myanmar National Building Code, 2016 [1] which is adapted from ASCE 7-05 [2] is stipulated to examine the analysis by the Federation of Myanmar Engineering Society (Fed. MES). As for the ground motion definition, the geometric mean spectrum is the most widely used horizontal definition to combine the two orthogonal spectral accelerations into a single numerical value [3]. The definition is also used in MNBC, 2016 and the framework is also similar to the National Earthquake Hazards Reduction Program (NEHRP), 2003 Seismic Design Provision [4]. Nowadays, the updated version of seismic provision and ASCE 7-16 [5] provide the maximum direction ground motion instead of geometric mean of two orthogonal ground motions components due to its short coming definition. The geomean definition does not give the ground motion values which vary with the orientation of recording axes. Thus, researchers developed the definition of median spectral acceleration (Sa_{RotD50}) and maximum spectral acceleration ($Sa_{RotD100}$) overall orientation [6]. In this study, the factors developed by Shahi and Baker, 2013 [7] which are the ratio of maximum spectral acceleration ($Sa_{RotD100}$) to median

IOP Conf. Series: Materials Science and Engineering 1153 (2021) 012001 doi:10.1088/1757-899X/1153/1/012001

spectral acceleration (Sa_{RotD50}) at discrete period is used to predict the maximum spectral acceleration overall orientations. In time history analysis, the different responses of a structure could be resulted by the different ground motions. There are three orthogonal components of ground motion that accelerogram records: two in the horizontal direction and one in vertical direction. In this study, the two-dimensional intensity of ground motion is taken into account to compute response spectrum by using two orthogonal horizontal components of the ground motion. The research objective of the study is to figure out how these two-horizontal definitions; regional code ground motion definition and the maximum ground motion effects on the reinforce concrete moment frame structure. The study location is in Mandalay, Myanmar which is very close to the major strike slip fault and equivalent lateral force analysis (EQLF), response spectrum analysis (RSA) and linear response history analysis (LRHA) are performed according to the MNBC, 2016. This research work is an attempt to reach on more accurate conclusion to reduce their effect on the structure.

2. Literature Review

A literature review states that related to the characteristic of ground motions, and seismicity of Mandalay City, Myanmar. A brief introduction about the ground motion characteristics, horizontal definitions of ground motion and overview of seismic hazard level for the study location are discussed in this section.

2.1 Characteristics of Ground Motions

In earthquake engineering, ground motion referred to as the ground acceleration, velocity and displacement. There are three parameters in terms of amplitude: peak ground acceleration (PGA), peak ground velocity (PGV) and peak ground displacement (PGD). They are the particular interests in most cases amplified when transmitted through the structure by ground vibration. The strong motion instruments are used to measure the earthquake ground motion and record the ground acceleration [8]. The recorded acceleration consists of two horizontal components and one in vertical direction. There are many ways to describe the earthquake ground motion intensity measures. 5% damped spectral acceleration can also be described in either single arbitrary components or the combination of two orthogonal horizontal components. The geometric mean response spectrum which is the combination of two horizontal orthogonal ground motion components is the most widely used horizontal definition and has been used in ASCE 7-05 and attenuation relationship. But the spectral acceleration of geomean depends on the sensor recording orientation. Thus, researchers proposed many others orientation independent definitions: $Sa_{GMRotDnn}$ and $Sa_{GMRotInn}$ by Boore et al, 2006 [9] and Sa_{RotDnn} and Sa_{RotInn} by Boore, 2010 [6] to remove the orientation independence. The PEER Ground Motion Database introduced new spectral acceleration to remove the recorded orientation; Sa_{RotD50} (median spectral acceleration) and $Sa_{RotD100}$ (maximum spectral acceleration) as an intensity measure. Current ground motion prediction equations (GMPE) use median spectral acceleration (Sa_{RotD50}) which is the 84th percentile of the geometric mean response [10]. Shahi and Baker developed the model for maximum direction spectral acceleration ($Sa_{RotD100}$) by using over 3000 ground motions from the NGA West-2 Database. The multiplicative factor which is the ratio of $Sa_{RotD100}$ to Sa_{RotD50} is evaluated to predict the maximum spectral acceleration of a model at a site. Shahi and Baker (2013) proposed that the ground motion directionality referred to as the polarization of the ground motion [10]. This leads to the discrepancy between response spectra definitions. The unpolarized case, e.g., "1994 Chi Chi earthquake" is defined when the ground motion has an approximately equal response spectrum over all orientations at a given period and the value of $Sa_{RotD100}$ to Sa_{RotD50} ratio is nearly 1. Nevertheless, the value of the ratios is equal to 1.414 if the ground motion is in strongly polarized case by the "1984 Morgan Hill earthquake". Thus, the value of Sa for any ground motion will be within the range of 1 to 1.414. [7]. According to Huang et al, 2007, the spectral demand by geomean ground motion can be much smaller than the maximum demand in the near fault region. The research is emphasized on the seismic records which are selected with seismic parameters of 6.5 moment magnitude or larger and less than 15 km site to source distance. The resulted earthquakes ground motion by next attenuation relationships gives maximum direction response spectral acceleration of the 110% of 5% damped at short period and 130% at 1 second period. [11].

2.2 Seismicity of Myanmar
Myanmar is one of the earthquake-prone zones in Southeast Asia. In this study, Mandalay City, Myanmar is selected for the site consideration. It is the second largest city and covers approximately from latitude 21° 51′ to 22° 1′ N and Longitude 96° 3′ to 96° 8′ E. It is located very close to the most active dextral Sagaing fault in Myanmar which is a major north-trending right lateral strike-slip fault and 8 km away from Mandalay City [12]. Mandalay City has a population around 1.3 million, according to the 2014 records and grow gradually. Thus, the scale of disaster from seismic excitation will be increased in the cities due to urbanization the vulnerability is increased. MNBC, 2016 brings together the information on the ground motion parameters which are described by the maximum considered earthquake (MCE) spectral response acceleration with 5% damping ratio at short period, S_s (0.2 sec) and at one second period, S_1 (1.0 sec) of Mandalay City are 2.01g and 0.8g respectively [1]. According to the study of the seismic microzonation report of Mandalay City, the spatial variation of V_{S30} ranges from 220 m/s to 340 m/s for nearly all of the sites in Mandalay [13]. Thus, the site class D (stiff soil) is defined for this V_{S30} range [14]. The response spectrum at design and maximum considered earthquake levels for both geomean and maximum spectral accelerations are illustrated in figure 1. The geomean code spectra for DBE and MCE are constructed by applying spectral acceleration parameters provided by MNBC, 2016 while the DBE and MCE of $Sa_{RotD100}$ spectra are constructed by using factors of Shahi and Baker, 2013 [7].

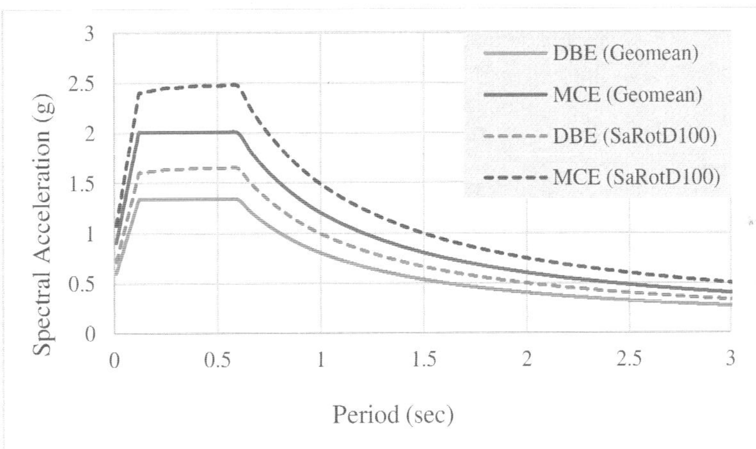

Figure. 1 MCE and DBE of geomean and $Sa_{RotD100}$ target spectra for Mandalay City.

3. Research Methodology
The study is carried out to analyze the behaviour of soft story rectangular building, provided as 4m base height and 3m typical floor height. The model of the building is created in ETABS software [15] and building structural properties are defined as table 1 according to the Myanmar National Building Code, 2016. Figure 2 shows the plan, elevation and 3D views of the building. Moreover, the effective stiffness of the structural elements such as beams, columns and slabs are taken into account according to the American Concrete Institute Code, 2005 [16]. Since Mandalay City, Myanmar is selected as the study site location, the seismic category is considered as E and soil type is stiff soil according to the Thein et. al, 2018. The applied loads such as super dead loads, live loads and earthquake loads to the office building structures are according to the MNBC, 2016 part III and Part IV respectively in table 2.

The seismic parameters used in this study are shown in the table 3 for the inelastic behaviour of the structure. The analysis is carried out by equivalent lateral force analysis, response spectrum analysis and linear response history analysis and determined the story displacement, story drift, story overturning moment and story shear. In this study, the designed base earthquake spectra and maximum considered earthquake spectra for both geomean and $Sa_{RotD100}$ is illustrated according to the regional code and by using the multiplicative factors developed by Shahi and Baker, 2013 [7]. The recorded earthquake data are obtained from NGA West-2 PEER Ground Motion Database [17] to perform linear response history analysis. After the analyses are done, the results obtained from ETABS software are turn observed to form conclusions.

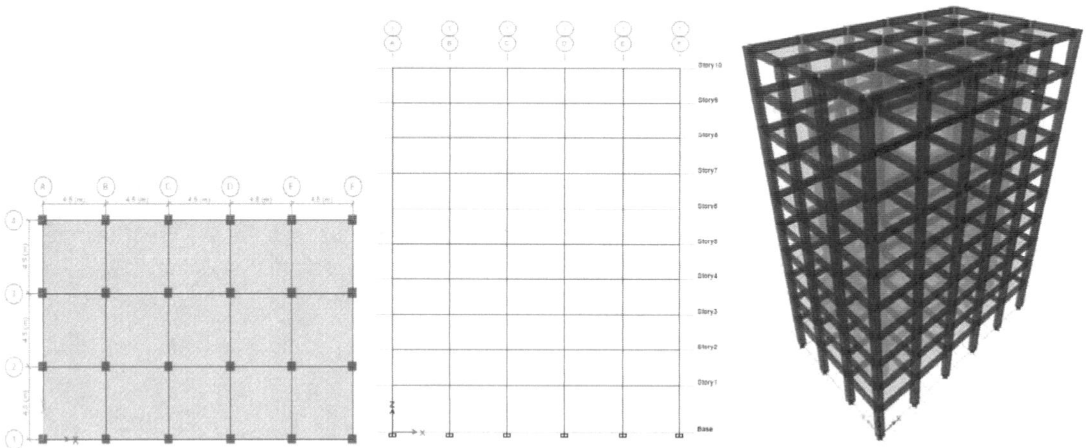

Figure 2. Plan, elevation and 3D views of model.

Table 1. Description of the building

Building Type	(Office) Special Concrete Moment Frame
No. of Story	10
Bay size	4.5m
Height	4m (Base) 3m (Typical)
Beam Dimension	350mm*500mm
Column Dimension	600mm*600mm
Slab Depth	130mm
f_c'	28 MPa
f_y	413 MPa

Table 2. Load assignment (MNBC, 2016)

Super Dead Load	1.5kN/m^2 (Typical) 1kN/m^2 (Roof)
Live Load	2.4kN/m^2 (Typical) 1kN/m^2 (Roof)

Table 3. Seismic parameters used in this study (MNBC, 2016)

Parameter	Value
S_s	2.01
S_1	0.8
S_{DS}	1.34
T_L	6
R	8
Ω	3
C_d	5.5
I	1

3.1 Analytical Procedures

In this study, permitted analytical procedures according to the MNBC, 2016 [1] are carried out to examine the structural responses and the analysis methods are as follows:

- Equivalent Lateral Force Analysis (EQLF)
- Response Spectrum Analysis (RSA)
- Linear Response History Analysis (LRHA)

Equivalent lateral force analysis is carried out to estimate the seismic design forces at any level. In this analysis, the induced base shear of the structure due to seismic excitation is calculated and it depends on the soil condition of the site, potential sources probability of seismic activity, structural properties, the total weight of the structure which includes 100% of the self-weight of the structural components, permanent weight and 25% of floor live loads. Moreover, the fundamental or natural period of the structure is also a basic factor which effects the seismic base shear. The seismic force distribution along the 10 -story building is calculated manually and compared with the ETABS results and presented in figure 3.

Response spectrum analysis is similar to the equivalent lateral force analysis. Nevertheless, the response spectrum analysis should be taken into account the sufficient number of modes while the equivalent lateral force analysis is only taken into account for a fundamental period. According to MNBC, 2016, at least 90% of the modal participation mass ratio are needed to be considered in each of the orthogonal directions of the building. Moreover, the base shear of the response spectrum analysis should be rescaled to reach a minimum of 85% equivalent lateral force base shear in modal analysis. The modal participation mass ratio of the first three modes is shown in the table 4.

Linear response history analysis is performed to determine the seismic structural responses which vary with the time. The ground motion data are needed to execute the linear response history analysis. These time histories ground motion data which are the acceleration versus time input signals are selected from NGA West-2 Ground Motion Database Web and scaled according to the regional Code.

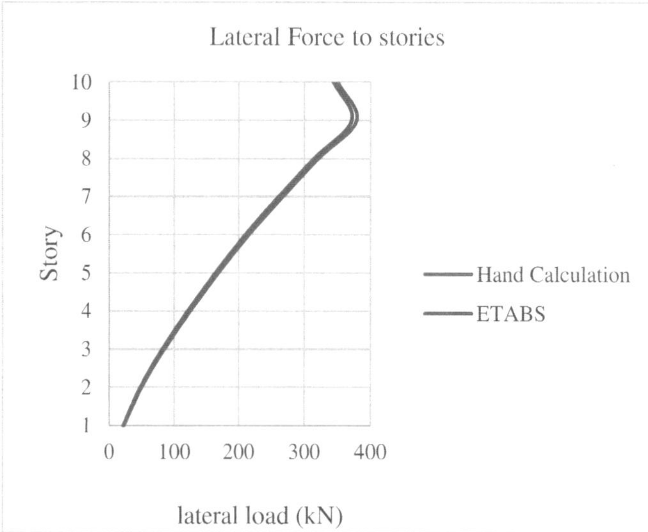

Figure 3. Lateral force to each story.

Table 4. Modal participation mass ratios

Modal Participating Mass Ratios				
Mode	Period (sec)	U_X	U_Y	R_Z
1	1.615	0	0.8143	0
2	1.549	0.8183	0	0
3	1.345	0	0	0.8197

3.2 Selection and Scaling of Ground Motions

In dynamic analysis, the adequate selection of ground motion becomes one of the important issues. The different ground motions can give the different structural responses. In this study, the ground motion records selection is done by using PEER West-2 ground motion database to perform linear response history analysis. The ground motion time histories are generated which have the similar ground motion characteristics as the elastic response spectra. According to MNBC, 2016, at least three ground motion time histories are needed for maximum response and average response is resulted by seven or more time histories. In this case, seven earthquake ground motions are selected to perform time history analysis according to Tall Building Initiative Guidelines, 2010 [18]. The PEER Ground Motion Database Web needs selection criteria such as tectonic type, magnitude, site to source distance, soil condition of specific site, frequency, scale factor limitation, spectral shape, maximum number of ground motion, considered period range to select the time history ground motions. The criteria to select the ground motions are as shown in the table 5. After that, the ground motion database obtained from the PEER website are scaled by mean square error within the period range of 0.2T to 1.5T and cooperated in ETABS according to the current code. The seven ground motion records resulted from the PEER Ground Motion Database and scale factors used for the analysis are shown in table 6 and table 7 for geomean response spectrum and $Sa_{RotD100}$ response spectrum respectively. Moreover, a set of seven bi-directional ground motion excitations for both geomean and maximum response spectra is chosen for three-dimensional analysis according to the code that is illustrated in figure 4.

Table 5. Ground motion selection criteria

Selection Criteria	Range
Magnitude	5.5-8
Fault Type	Strike-slip
Closest Distance to Fault Rupture Plane	0.1-15 (km)
V_{S30}	200-400(m/s)
Intensity Measures	Geomean $Sa_{RotD100}$
Damping Ratio	5%
Suite Average	Geometric Mean
Scale Factor	1-4

Table 6. Selected earthquake ground motions for geomean response spectra (PEER, 2010)

No	RSN	Earthquake Name	Year	Station Name	M_w	R_{jb} km	R_{rup} km	V_{s30} m/sec	SF
1	6	Imperial Valley-02	1940	El Centro Array #9	6.95	6.09	6.09	213.44	3.4837
2	165	Imperial Valley-06	1979	Chihuahua	6.53	7.29	7.29	242.05	3.6564
3	183	Imperial Valley-06	1979	El Centro Array #8	6.53	3.86	3.86	206.08	2.8115
4	864	Landers	1992	Joshua Tree	7.28	11.03	11.03	379.32	2.9455
5	1101	Kobe_ Japan	1995	Amagasaki	6.9	11.34	11.34	256	2.1305
6	5825	El Mayor-Cucapah_Mexico	2010	CERRO PRIETO GEOTHERMAL	7.2	8.88	10.92	242.05	2.8933
7	6893	Darfield_ New Zealand	2010	DFHS	7	11.86	11.86	344.02	3.4545

Table 7. Selected earthquake ground motions for $Sa_{RotD100}$ response spectra (PEER, 2010)

No	RSN	Earthquake Name	Year	Station Name	M_w	R_{jb} km	R_{rup} km	V_{s30} m/sec	SF
1	183	Imperial Valley-06	1979	El Centro Array #8	6.53	3.86	3.86	206.08	3.4743
2	558	Chalfant Valley-02	1986	Zack Brothers Ranch	6.19	6.44	7.58	316.19	2.975
3	725	Superstition Hills-02	1987	Poe Road (temp)	6.54	11.16	11.16	316.64	3.9365
4	864	Landers	1992	Joshua Tree	7.28	11.03	11.03	379.32	3.6399
5	1101	Kobe_ Japan	1995	Amagasaki	6.9	11.34	11.34	256	2.6328
6	5825	El Mayor-Cucapah_Mexico	2010	CERRO PRIETO GEOTHERMAL	7.2	8.88	10.92	242.05	3.5753
7	5829	El Mayor-Cucapah_Mexico	2010	RIITO	7.2	13.7	13.71	242.05	3.0943

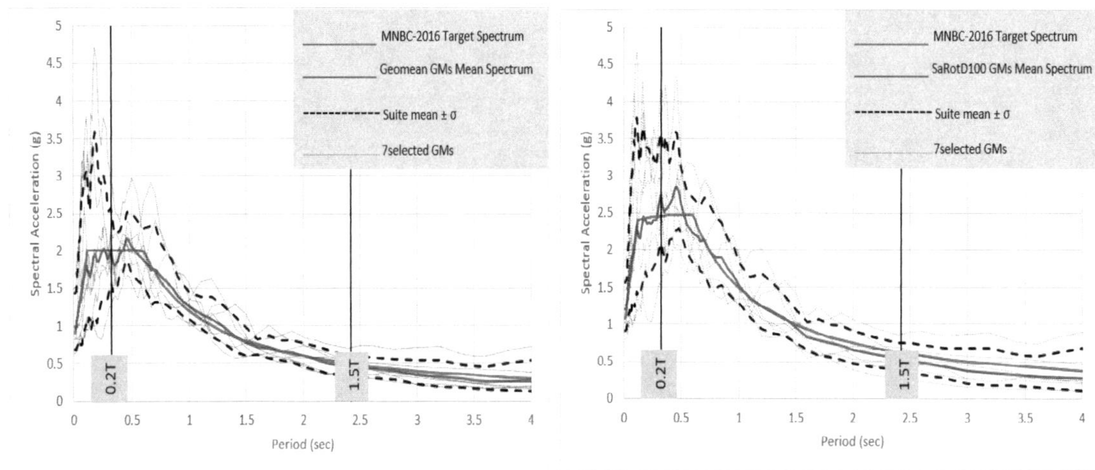

Figure 4. Geomean ground motion set and $Sa_{RotD100}$ ground motions set.

4. Results and Discussion

The story displacement, story drift, story shear and story overturning moment are analyzed in ETABS, 2016 by using Equivalent lateral force analysis (EQLF), response spectrum analysis (RSA) and linear response history analysis (LRHA). These three analyses are conducted according to MNBC, 2016 to investigate the seismic structural responses. The linear response history of three-dimensional analysis is responsible by bi-directional excitation of each ground motion of the geometric mean directionality combination and maximum directionality of ground motions. The two spectra at maximum considered earthquake level are employed as target spectra to select seven ground motions from the PEER Ground Motion Database for linear response history analysis. The average responses of both geomean and maximum direction response spectra in each direction (X and Y directions) are calculated and illustrated in the following figure 5-8. The story drift increases from bottom story to the 3rd story and gradually decreases for all analyses. It can be observed that the maximum story drift occurs by $Sa_{RotD100}$ in LRHA and RSA results are the lowest in both X and Y direction. The story displacement in EQLF analysis is different around 5-10% from X and Y direction due to its change of moment of inertia. The geomean LTHA overturning moment in Y-direction is slightly lower than that of EQLF and higher above 4th story in both structures. The maximum difference of 30% and the minimum value is 10% between $Sa_{RotD100}$ and geomean definition in LTHA. The story shear in X-direction all over analyses have the same pattern with other responses and have lower difference in the mid-story levels in Y-direction. The ratio between required strength and structural elasticity, termed as response modification factor is taken as 8 in all analyses for elastic demands. The base shear of the RSA is scaled to reach the 85% of EQLF according to the code. Thus, the contribution of structural response in EQLF overestimates than that of RSA because it ignores the higher mode of the structures. Since the response spectrum of $Sa_{RotD100}$ has larger values than the geomean response spectrum, the selected ground motions also have higher spectral acceleration values for $Sa_{RotD100}$. Based on Huang et al 2008, maximum response spectral acceleration at short and 1 second period have larger values than that geomean response spectral acceleration. Thus, the larger story responses are resulted by $Sa_{RotD100}$ compared with the geometric mean time history analysis in this study. As a result, approximate overall range of 10%-30% difference is observed in both directions by LRHA for geomean and maximum direction response spectra. Thus, the amount of these differences is significant in a performing seismic analysis.

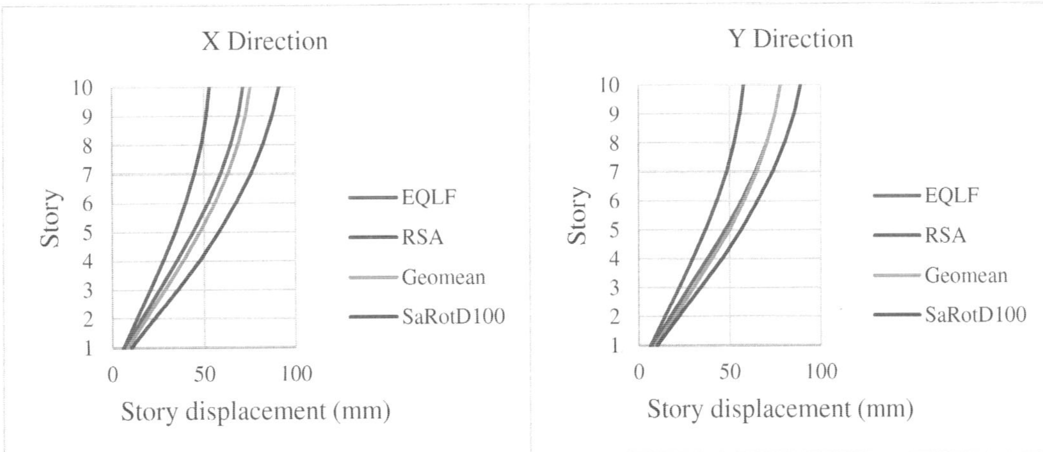

Figure 5. Story displacement from EQLF, RSA and LRHA (geomean and $Sa_{RotD100}$).

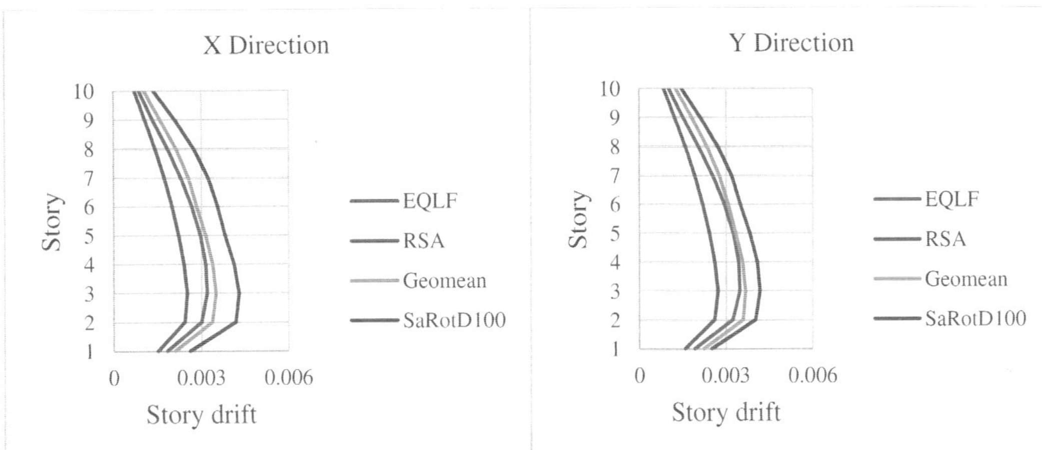

Figure 6. Story drift from EQLF, RSA and LRHA (geomean and $Sa_{RotD100}$).

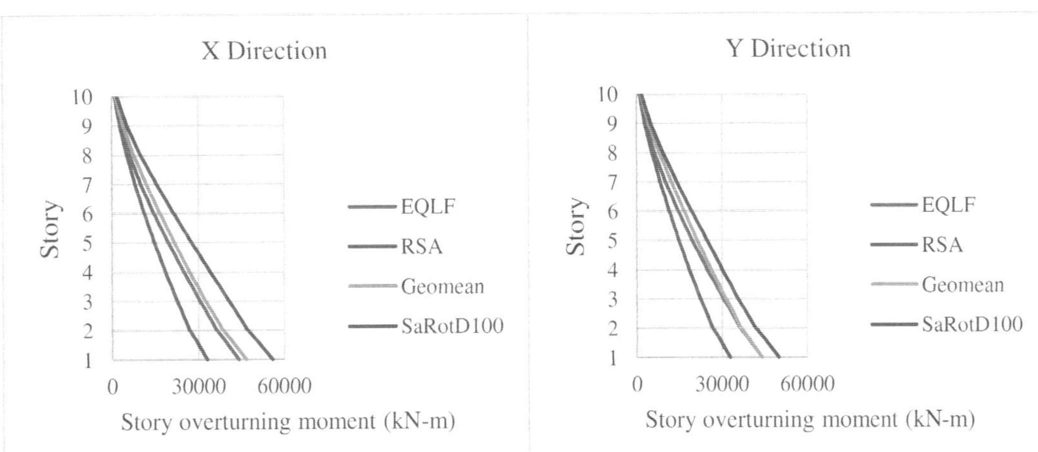

Figure 7. Story overturning moment from EQLF, RSA and LRHA (geomean and $Sa_{RotD100}$).

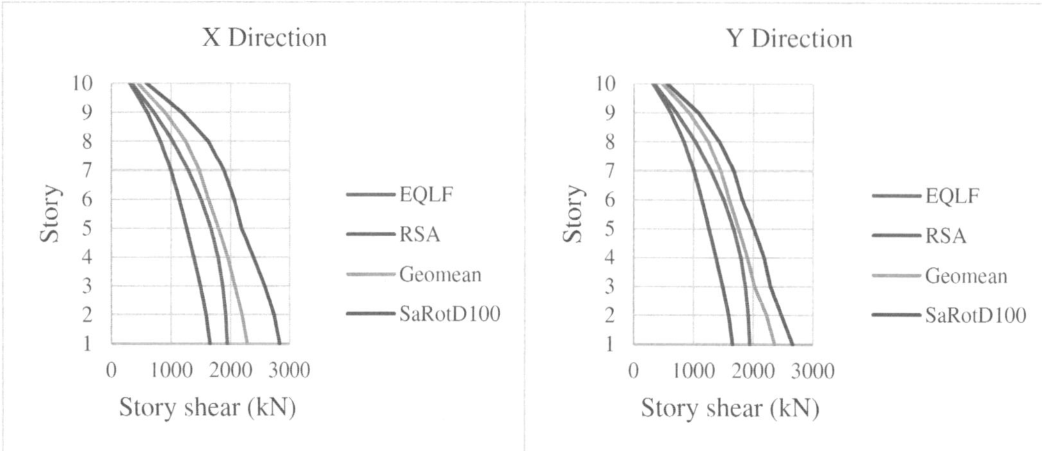

Figure 8. Story shear from EQLF, RSA and LRHA (geomean and $Sa_{RotD100}$).

5. Conclusion

In this study, the effects of two horizontal definitions in ground motion selection are investigated on soft story rectangular building which is located in Mandalay city, Myanmar. Both static and dynamic analyses such as equivalent lateral force analysis, response spectrum analysis and linear response history analysis have been done to examine the seismic demand of the structures by geomean which is currently used in the Myanmar National Building Code, 2016 and $Sa_{RotD100}$ which will be used in future. The different ground motions are obtained by different MCE spectra in the selection of ground motions. In general, larger structural responses are controlled by the maximum direction definition compared to the geomean horizontal definition. The results showed the considerable effect around 10% to 30% on the response of the structure by maximum direction spectral accelerations even though MNBC, 2016 counts for geomean spectral acceleration. On the other hand, using more ground motions could capture all the variability of different characteristics of every single ground motion. Thus, it gave the option to consider the maximum response acceleration in ground motion selection and to reduce the effect of the structure in MNBC code. The consideration of maximum effect may not be economically feasible compared to the current code in Myanmar. Nevertheless, the structures should be designed in proper ways because the effect of the earthquakes may cause the structures collapse and life of people may spoil.

References

[1] Myanmar Engineering Council (2016). *Myanmar national building code.*

[2] ASCE. (2005). ASCE standard, minimum design loads for buildings and other structures.

[3] Beyer, K. and Bommer, J.J. (2006). Relationships between median values and between aleatory variabilities for different definitions of the horizontal component of motion. *Bulletin of the Seismological Society of America*, *96*(4A), pp.1512-1522.

[4] Federal Emergency Management Agency. (2003). NEHRP recommended provisions for seismic regulations for new buildings and other structures. Fema.

[5] ASCE/SEI 7-16 (2016). Minimum Design Loads for Buildings and Other Structures, American Society of Civil Engineers, Reston, VA.

[6] Boore, D.M. (2010). Orientation-independent, nongeometric-mean measures of seismic intensity from two horizontal components of motion. *Bulletin of the Seismological Society of America*, *100*(4), pp.1830-1835.

[7] Shahi, S.K. and Baker, J.W. (2014). NGA-West2 models for ground motion directionality. *Earthquake Spectra*, *30*(3), pp.1285-1300.

IOP Conf. Series: Materials Science and Engineering 1153 (2021) 012001 doi:10.1088/1757-899X/1153/1/012001

[8] Mayes, R. L., & Naeim, F. (2001). The Seismic Design Handbook 2nd Edition Ch. 14 Design of Structures with Seismic Isolation.

[9] Boore, D.M., Watson-Lamprey, J. and Abrahamson, N.A. (2006). Orientation-independent measures of ground motion. *Bulletin of the seismological Society of America*, *96*(4A), pp.1502-1511.

[10] Luco, N., Bachman, R.E., Crouse, C.B., Harris, J.R., Hooper, J.D., Kircher, C.A., Caldwell, P.J. and Rukstales, K.S., 2015. Updates to building-code maps for the 2015 NEHRP recommended seismic provisions. *Earthquake Spectra*, *31*(1_suppl), pp.S245-S271.

[11] Huang, Y.N., Whittaker, A.S. and Luco, N. (2008). Maximum spectral demands in the near-fault region. *Earthquake Spectra*, *24*(1), pp.319-341.

[12] Thein M, Swe TL, Han S. The Seismic Zone Map of Myanmar. (2006).

[13] Thein, P.S., Kiyono, J., Win, T.T., Nu, T.T. and Aung, D.W. (2018). Seismic Microzonation of Mandalay City, Myanmar. *Journal of Geological Resource and Engineering*, *6*, pp.1-13.

[14] Nwe ZZ, Tun KT. Seismic Hazard Analysis using AHP-GIS. Int. J. Res. Chem. Metallurg. Civ. Eng. 2016;3:1442-50.

[15] CSI, C. (2016). Analysis reference manual for SAP2000, ETABS, and SAFE. Computers and Structures, Inc., Berkeley, California, USA.

[16] ACI Committee. (2005). Building code requirements for structural concrete (ACI 318-05) and commentary (ACI 318R-05). American Concrete Institute.

[17] Haque, M., Ray, S., Chakraborty, A., Elias, M. and Alam, I. (2016). Seismic Performance Analysis of RCC Multi-Storied Buildings with Plan Irregularity. *American Journal of Civil Engineering*, *4*(2), pp.52-57.

[18] Pacific Earthquake Engineering Research Center. (2010). *Guidelines for performance-based seismic design of tall buildings*. Pacific Earthquake Engineering Research Center, College of Engineering, University of California.

[19] TBI Guidelines Working Group. Guidelines for Performance-Based Seismic Design of Tall Buildings, Berkeley, California: Pacific Earthquake Engineering Research Center, 2010.

Reliability-based prediction of landslide using efficient Monte Carlo Simulation

E J Joe, N Ali and S N Jusoh

School of Civil Engineering, Faculty of Engineering, Universiti Teknologi Malaysia, 81310 Skudai, Johor, Malaysia
E-mail: edgarjr.joe@gmail.com

Abstract. Several studies have attempted to develop thresholds for prediction of landslide initiation due to antecedent rainfall by using the physical-based approach. The approach offers a better understanding of the underlying process of the landslide mechanism. In the previous studies, however, determinate values of hydro-mechanical soil parameters are commonly considered, neglecting the uncertainties of the variables. Thus, this study aims to propose a relatively computationally, inexpensive reliability approach to address the uncertainties of the multivariate. The application of the proposed model is demonstrated by taking a case study from Sabah, Malaysia. The Latin Hypercube Sampling was adopted for efficient sampling, while the dependencies of the multivariate were modelled using a higher-order copula. The infiltration process and slope stability were assessed based on a fully-coupled seepage-deformation and shear strength reduction technique, respectively in a finite element programme. The probabilities of failure for various antecedent rainfall condition were computed by using Monte Carlo Simulation. In order to reduce the computational cost of the simulation, the Multilayer Perceptron regressor was applied to form the non-linear surrogate model for an implicit limit state function.

1. Introduction

In the tropical climate region, many rainfall-induced landslides have been reported, such as those in China, India, Indonesia, Singapore, Thailand, as well as in Malaysia. The occurrences of the landslide can be predicted by using the physical-based threshold [1]. The physical-based (also known as conceptual or process-based) threshold involves a simulation of landslide mechanism due to rainfall infiltration. In unsaturated residual soil, Rahardjo *et al.* [2] have highlighted the significant role of antecedent rainfall in triggering a landslide. The antecedent rainfall can be defined as the precipitation that occurs in several days closely before the landslide. Their conclusion concurs with the previous study by Wei *et al.* [3] who have found out that the landslide at Bukit Batok, Singapore happened after several days of torrential downpour, but no rainfall occurred during the landslide incident. Their findings conclude that a landslide is not necessarily triggered by rainfall on the landslide event and provide evidence of the critical role of antecedent rainfall in triggering a landslide.

In the development of the physical-based threshold, most of the previous researches considered determinate value for each hydraulic and mechanical soil parameters, except that by Dou *et al.* [4] and Ering and Babu [5] to the best of the author's knowledge. However, the geomaterial contains a certain degree of uncertainties due to spatial variability, random testing error, sub-standard measurement procedure, and statistical error such as too few data [6]. Dou *et al.* [4] applied the Monte Carlo Simulation by considering the variability in hydraulic conductivity for a hypothetical infinite slope. Ering and Babu [5] considered the uncertainties of cohesion, friction angle, and matric suction by adopting the Bayesian analysis for a case study in Malin, India. In the Bayesian approach, the parameter value which characterise a system is inferred based on prior information of measured value. However, the approach suffers from the main limitation that the selection of prior probability is subjective and requires information from previous studies, which may not be available or inadequate. In the broader context of the rainfall-induced landslide assessment, several other studies have adopted reliability analysis [7, 8]. Nonetheless, the construction of physical threshold for landslide initiation due to antecedent rainfall, which consider the various uncertainties in hydromechanical soil parameters is still lacking.

Therefore, this study aims to propose a relatively computationally, inexpensive reliability analysis framework to construct the physical threshold for the landslide initiation in unsaturated soil, subjected to antecedent rainfall. The uncertainties of random variables of both hydraulic and mechanical soil parameters are considered. The application of the proposed model will be demonstrated based on a case study of a rainfall-induced landslide in Sabah, Malaysia.

2. Study area

The capital city of Sabah state in Malaysia, *i.e.* Kota Kinabalu is one of the densely populated areas in the state. One of the landslide-prone areas within the vicinity of Kota Kinabalu is situated around a local access road, known as Jalan Penempatan (Figure 1).

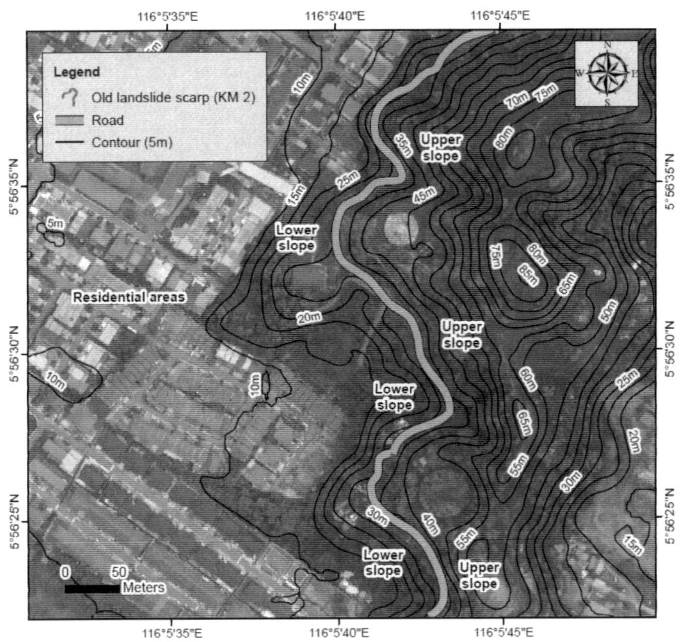

Figure 1. Location map of study area.

The road which stretches of about 3 km long (from 5.954484 °N, 116.094497 °E to 5.934683 °N, 116.092300 °E) serves as an important route which connects several housing areas to the city centre. The study location which covers an area of about 1.5 km^2, pertains to a natural hilly terrain with an elevation from 10 to 85 m above mean sea level. The site is underlain by the Eocene-Oligocene Crocker Formation, which mainly consisted of sandstone and shale. The residual soil typically consists of stiff to very hard sandy, clayey silt with an average depth of 10 m. The bedrock comprised of very weak to moderately strong sandstone.

A major landslide in the study area has occurred on 9th June 2010 at Kilometre 2 (5.941728 °N, 116.094071 °E). The landslide event was due to occurrence of about 9-day rainfall before the incident and a major storm on the landslide day itself (Figure 2), which happen from 31st May 2010 until 9th June 2010. However, no rainfall was recorded on fifth (4th June 2010) and eighth day (1st June 2010) before the event. The cumulative rainfall for the ten days was 346.8 mm, with the maximum rainfall of 74 mm during the landslide day.

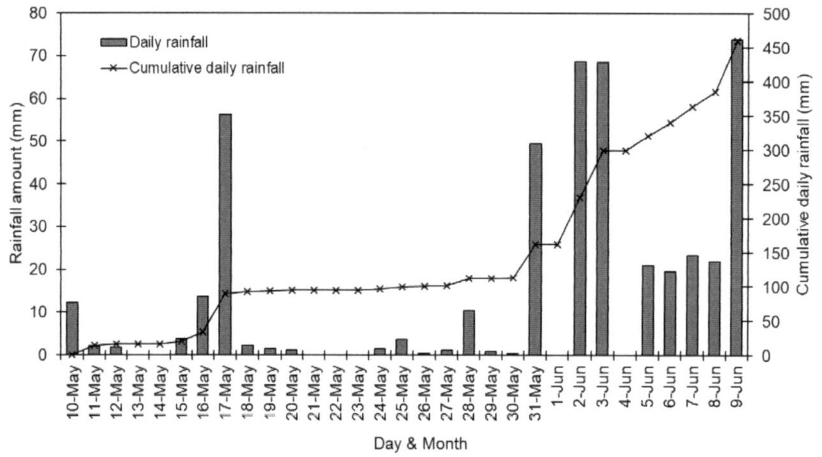

Figure 2. Rainfall record at Kota Kinabalu rainfall station in a month before landslide event on 9th June 2010.

3. Methodology

3.1. Coupled hydro-mechanical constitutive model

3.1.1. Fundamental flow equation. In this study, the constitutive behaviour of pore fluid flow in the soil is governed by Darcy's law. In a transient state, the seepage through unsaturated soils based on Darcy's law can be written in terms of mass conservation as shown in Equation 1.

$$\nabla^T[K(\theta)\nabla(\Phi + z)] = c(\theta)\frac{\partial \Phi}{\partial t} \tag{1}$$

where $K(\theta)$ is permeability; Φ is pressure head; z is elevation head; t is time, and; $c(\theta)$ is specific moisture capacity.

3.1.2 Hydraulic model. The Soil Water Characteristic Curve (SWCC) was fitted using the two-parameter equation by Van Genuchten [8] as shown in Equation 2.

$$\theta_\psi = \theta_r + \frac{\theta_s - \theta_r}{[1 + |\alpha\psi|^n]^{1-\frac{1}{n}}} \tag{2}$$

where θ_ψ is volumetric water content at a matric suction; θ_r is residual volumetric water content, θ_s is saturated volumetric water content; ψ is matric suction, and; α, n are fitting parameter. The fitting parameter α (also known as scaling parameter) is related to the air entry value of the SWCC, while the fitting parameter n (also known as the shape parameter) defines the slope of the SWCC. The best-fitting parameters can be determined by adopting the technique of curve fitting, in which the objective is set to minimise the sum of squared errors between the measured volumetric water content and the modelled volumetric water content.

The combined equation by Van Genuchten [8] and Mualem [9] was used to derive the hydraulic conductivity function as shown in Equation 3.

$$K_\psi = K_s \times {S_e}^l \times [1 - (1 - S_e^{\frac{1}{m}})^m]^2 \tag{3}$$

where K_ψ is hydraulic conductivity at a matric suction, K_s is saturated hydraulic conductivity, S_e is effective water content, l is parameter describing the pore structure of soil, and; m is a fitting parameter calculated as $1-1/n$.

3.1.3 Elasto-plastic model. The shear strength of the unsaturated slope in this study was determined based on the modified Mohr-Coulomb failure criterion as shown in Equation 4.

$$\tau_f = c' + [(\sigma - u_a) + \chi(u_a - u_w)] \tan \phi' \quad (4)$$

where τ_f is the shear strength; c' is effective cohesion; ϕ' is effective internal friction angle; σ is total normal stress; u_a is pore air pressure; u_w is pore water pressure; and χ is an effective stress parameter which is a function of degree of saturation.

3.2. Geostatistical model and dependencies of multivariate

In this study, eight number of hydro-mechanical soil parameters namely the saturated hydraulic conductivity (k_s), fitting parameter α and n of SWCC, saturated volumetric water content (θ_s), residual volumetric water content (θ_r), unit weight (γ), effective cohesion (c'), and effective internal friction angle (ϕ') were considered as the random variables. The geostatistical modelling involved the best fitting of marginal distributions and the generation of random samples. Based on these data, the dependencies of the multivariate can be constructed by using a higher-order copula.

The best-fit marginal distribution for each variable was identified based on the lowest score of Akaike Information Criterion (AIC) and Bayesian Information Criterion (BIC) by applying Equation 5 and 6. Four candidates of marginal distribution, *i.e.* normal, lognormal, Gumbel, and Weibull distributions were considered.

$$AIC = -2\sum_{i=1}^{N} ln(f(x_i)) + 2k \quad (5)$$

$$BIC = -2\sum_{i=1}^{N} ln(f(x_i)) + k\, log(N) \quad (6)$$

where N is the number of input variable; $f(x_i)$ is the distribution of variable x_i; and k is the number of distribution parameter.

Then, the Latin Hypercube Sampling was applied for efficient sampling, where the space of sample was evenly divided into possible subspaces, and samples point in each subspace were selected concurrently for an equal density of sampling in each subspace. No specific guideline is available on the required number of sample size. However, Silvestrini [10] have proposed sample size of 15D, where D is the dimension of input variables, as a safe rule of thumb for effective computer experiment with acceptable accuracy. Thus, in this study, the number of samples was taken as 15D to maintain both accuracy and computational cost.

Next, the dependencies of multivariate amongst the eight soil parameters were modelled using the drawable vine (D-vine) copula. The vine copula allows the modelling of the relationship amongst higher-order variables by taking the bivariate copula as the building block. Besides, different types of bivariate copula can be selected for the vine copula construction. In this study, five models of bivariate copula, namely the Clayton, Frank, Gumbel, Gaussian, and Student's t were considered as the candidates of the best-fit bivariate copula. The D-vine copula can be constructed according to Equation 7.

$$f(x_1, \dots, x_d) = \prod_{k=1}^{d} f_k(x_k) \times \prod_{i=1}^{d-1} \prod_{e \in E_i} c_{j(e),k(e)|D(e)} \left(F\left(x_{j(e)}|x_{D(e)}\right), F\left(x_{k(e)}|x_{D(e)}\right)\right) \quad (7)$$

where $f_k, k = 1, \dots, d,$ represent the marginal densities and $c_{i,i+j|1:(i-1)}$ bivariate copula densities with parameter(s) $\theta_{i,i+j|1:(i-1)}$ for j-th tree and i-th edge on tree T_j. The external product runs over the $d - 1$ trees and root nodes i, while the internal product denotes the $d - i$ pair copulas in each tree $i = 1, \dots, d - 1$.

For vector v of n-dimension, the functions of conditional distribution $F(x|v)$ may be calculated in term of pair copula for tree $n-1$, applying the pair-copulas of preceding trees 1, …, n. The relationship in Equation 8 is applied sequentially.

$$h(x|v,\theta) := F(x|v) = \frac{\delta C_{xv_j|v_{-j}}\left(F(x|v_{-j}), F(v_j|v_{-j})|\theta\right)}{\delta F(v_j|v_{-j})} \tag{8}$$

where v_j is a random element of v, and v_{-j} represents the $(n-1)$-dimensional vector v except v_j. The function of bivariate copula is defined by $C_{xv_j|v_{-j}}$ with parameter θ defined in tree n.

The most appropriate bivariate copula was identified based on the lowest scores of AIC and BIC (Equation 9 and 10).

$$AIC = -2\sum_{i=1}^{N} ln\, c\,(u_{1i}, u_{2i}; arg\, max\, L(\theta)\,) + 2k \tag{9}$$

$$BIC = -2\sum_{i=1}^{N} ln\, c\,(u_{1i}, u_{2i}; arg\, max\, L(\theta)\,) + k\log(N) \tag{10}$$

where θ is copula parameter, L is the log-likelihood function, u is standard uniform random samples, N is the number of input variable, and k is the unknown variable. For Clayton, Frank, Gaussian, and Gumbel copula, $\theta = 1$, and therefore, $k = 1$. For the Student's t copula, a degree-of-freedom, v is considered in addition to the copula parameter θ, and thus, $k = 2$.

Random samples from the cumulative distribution function, constructed from the conditional bivariate copulas can be generated by performing Rosenblatt Transformation. The transformation involves successive conditioning, as explained in Equation 11.

$$\begin{aligned} x_1 &= F_1^{-1}(u_1) \\ x_2 &= F_{2|1}^{-1}(u_2 \mid x_1) \\ &\vdots \\ x_i &= F_{i|1,2,\dots,(i-1)}^{-1}(u_i \mid x_1, x_2, \dots, x_{i-1}) \end{aligned} \tag{11}$$

where x_i is random samples, u is standard uniform random samples [0,1], and F_i^{-1} is inverse of the marginal cumulative distribution function.

3.3. Finite element model

A two-dimensional slope geometry (Figure 3) was modelled in a general-purpose finite element software, Abaqus. Homogeneous soil layer of sandy, clayey silt was considered in the simulation. The boundary conditions of the slope model were assigned as follows: (i) GH and EF as Dirichlet seepage indicated by the groundwater table; (ii) AH, DE, and FG as zero flux to develop the pore water pressure within the slope model; (iii) AB, BC, and CD as surface flux due to rainfall infiltration; (iv) ABCD as drainage-only condition so that developed pore pressure is not greater than zero; (v) AG and DF were restricted in the X-axis direction; and (vi) GF was restricted in both X- and Y-axes direction.

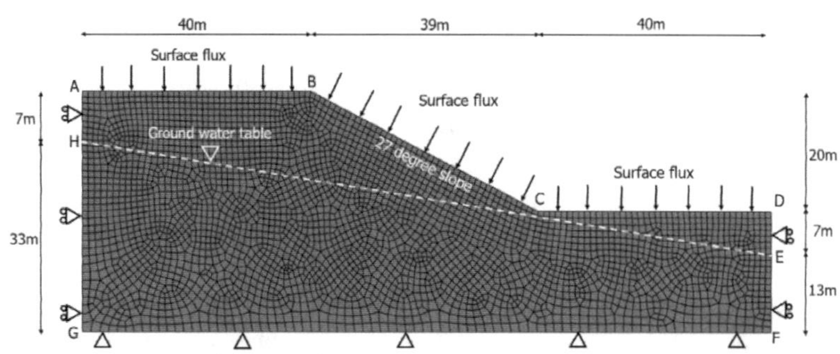

Figure 3. Slope mesh, geometry, and boundary conditions.

The element type used was a four-node plane strain quadrilateral with free mesh type. Mesh convergence study has been performed to arrive at an acceptable accuracy of finite element result with a reasonable computational cost. Therefore, the global mesh size of 1 m was applied on the slope model, while a localised coarser mesh of 1.5 m was applied at the base of the slope model to reduce the computational cost. After performing the coupled seepage-deformation analysis, the shear strength reduction technique was applied to compute the factor of safety.

3.4. Regression analysis based on machine learning

A regression analysis was performed to determine the limit state function of the sampling-based reliability analysis. In this study, the feed-forward artificial neural network of Multilayer Perceptron was selected for the supervised regression. The multilayer perceptron network consists of three layers, namely the input, hidden, and output layers. The number of hidden layer in this study was fixed at one as it is found that the single layer is adequate to train any complex non-linear functions. Not only will redundant hidden layers adding model complexity, but they also tend to cause overfitting of the model. The learning and convergence rate using a single hidden layer is also more flexible and faster, respectively, as compared to multiple hidden layers. The relatively latest method with improved performance, proposed by Sheela and Deepa [11] was adopted in this study as an initial guess on the required number of the hidden node. The method is explained by Equation 12.

$$N_h = \frac{4{n_i}^2 + 3}{{n_i}^2 - 8} \tag{12}$$

where N_h is the number of hidden node and n_i in the number of input node.

Firstly, the training and testing dataset were randomly split into 70% and 30% respectively. Then, the input datasets were normalised according to the minimum-maximum scaling. The early stopping method was applied to achieve an optimum model performance while preventing overfitting. Next, a 10-fold cross-validation was adopted for the training of the model to minimise the bias due to the random selection of training and remaining dataset. The performances from the 10-fold cross-validation were then averaged to assess the overall performance of the model training.

The relationship between the input vector x and output vector y in a Multilayer Perceptron network can generally be expressed by Equation 13.

$$y = f_{output}\left[\sum_{o=1}^{N_h} w_{ho}\, f_{hidden}\left(\sum_{h=1}^{n_i} w_{ih}\, x_i + b_h\right) + b_o\right] \tag{13}$$

where n_i is the number of input node; N_h is the node number in the hidden layer; x_i is the ith input unit; w_{ih} is the weight between the input node i and hidden node h; w_{ho} is the weight between the hidden node h and output node o; b_h is the bias (also known as threshold) for the hidden node h; b_o is the bias for the output node o; f_{hidden} is the transfer function of the hidden layer; and f_{output} is the transfer function for the output layer.

Lastly, the performances of the regression model for the training and testing dataset were assessed in terms of correlation coefficient (R^2), root mean squared error (RMSE), and mean absolute error (MAE).

3.5. Sampling-based reliability analysis

A sampling-based reliability analysis, namely the Monte Carlo Simulation (MCS), has been adopted in this study. The main idea of MCS is the generation of sets of random numbers according to the probability density function of the input variables and the performance function for each set are then computed. The MCS exhibits several merits such as a large number of simulations can be realised to achieve higher accuracy, a various statistical distribution of variables can be considered, the error in result can be easily determined, the velocity of simulation convergence is independent of the dimension of the random variables, the simulation procedure is not affected by the complexity in performance function, and the system failure probabilities of slopes which are governed by several representative slip surface can be captured [12]. However, the direct simulation demands a substantial computational cost, especially in the case of a smaller probability of failure (*e.g.* lesser than 10^{-6}). Therefore, the MCS is integrated with the regression function from the Multilayer Perceptron to address the processing time efficiently. In this study, it was found that the probability of failure and coefficient of variation were converging towards the 100,000-realisation number. The expected performance levels for the different probability of failure can be evaluated based on the target reliability indices published by the US Army Corps of Engineers [13].

4. Results and discussions

4.1. Geostatistical and multivariate dependencies model

Based on the lowest scores of AIC and BIC (Table 1), the saturated hydraulic conductivity, fitting parameter α and n of SWCC, saturated water content, and effective friction angle can be best-fitted by using the lognormal distribution. The unit weight and effective cohesion are best-fitted by the normal distribution. The Gumbel type is the most appropriate distribution to represent the residual water content.

Table 1. Summary of AIC and BIC for hydro-mechanical soil parameters at study area.

	Normal		Lognormal		Gumbel		Weibull	
	AIC	BIC	AIC	BIC	AIC	BIC	AIC	BIC
k_s	−58.4880	−55.8963	**−166.8142**	**−164.2225**	−28.4981	−25.9065	−159.2680	−156.6763
α	−110.8645	−108.2728	**−144.7285**	**−142.1368**	−86.1162	−83.5245	−136.6150	−134.0233
n	−3.3285	−0.7368	**−6.1193**	**−3.5277**	8.0464	10.6381	2.4490	5.0406
θ_s	−105.8611	−103.2694	**−106.0946**	**−103.5030**	−101.7371	−99.1455	−103.5229	−100.9312
θ_r	−90.8393	−89.5611	−41.4841	−40.2060	**−92.7105**	**−91.4324**	−88.0133	−86.7351
γ	**22.7893**	**23.7591**	22.8925	23.8623	23.5562	24.5260	23.3736	24.3434
c'	**57.6005**	**58.5703**	69.0684	70.0382	58.3290	59.2988	58.1741	59.1440
φ'	74.7262	75.6960	**73.8621**	**74.8319**	77.5225	78.4923	75.9846	76.9544

Note: Bold font indicates the best-fit marginal distribution type for each soil parameters.

Then, the dependencies of the multivariate were modelled by using the D-vine copula (Figure 4). At the first tree of the vine copula model, the sequence of pair of variables is determined by identifying the order with the maximum summation of the total pairwise Kendall's correlation coefficient. Thus, the bivariates in the first tree are arranged in the following sequence: 1) unit weight with saturated water content (joint by Clayton copula); 2) effective friction angle with unit weight (joint by Frank copula); 3) effective cohesion with effective friction angle (joint by Student's t distribution); 4) residual water content and effective cohesion (joint by Student's t distribution); 5) shape parameter n and residual water content (joint by Clayton copula); 6) saturated hydraulic conductivity and shape parameter n (joint by Clayton copula); and 7) scaling parameter α and saturated hydraulic conductivity (joint by Clayton copula). In tree two, six conditional bivariate copulas are constructed by the decomposition of the copula in the tree one. The decompositions of the conditional bivariate copulas in tree two were then performed to construct new conditional bivariate copulas in tree three. These steps were continued until the last tree. In the D-vine copula, the Clayton copula was applied for ten times, the Student's *t* copula was selected for eight times, the Frank copula was chosen for four times, and lastly, the Gaussian and Gumbel copulas were chosen for three times respectively. In total, 28 numbers of pair copula were constructed in the D-vine copula.

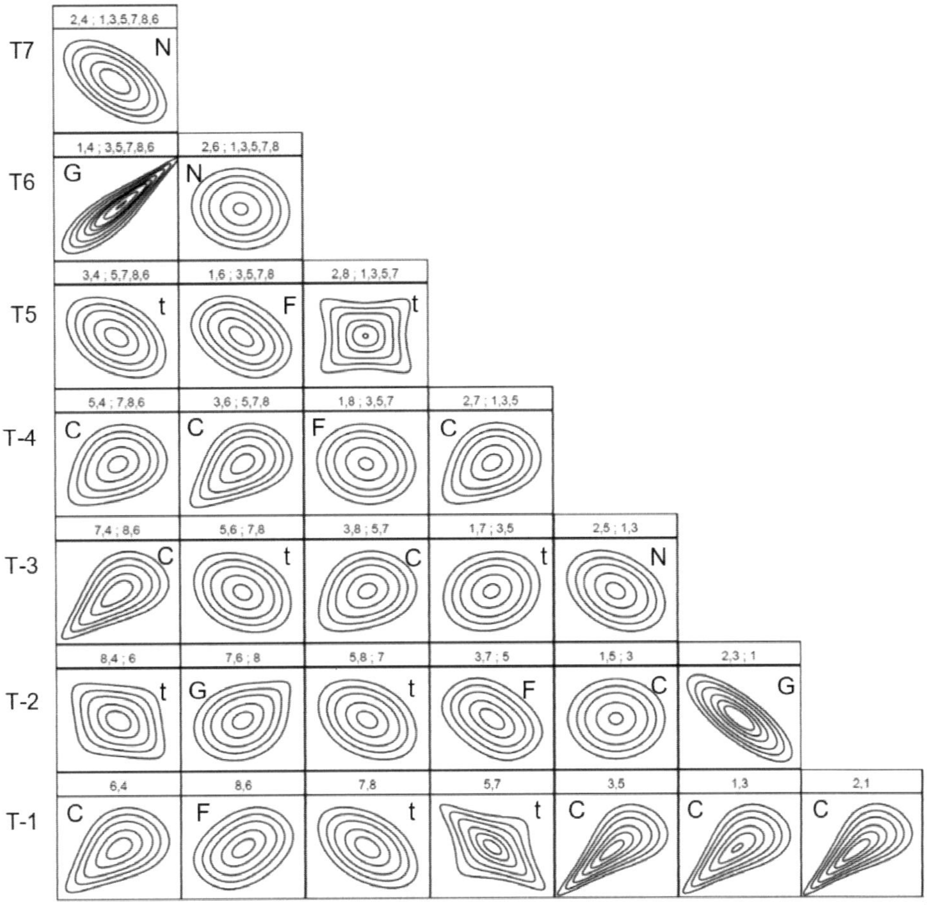

Figure 4. Conditional bivariate copula modelled for seven trees in the D-vine copula. The letter in upper-right or -left corner of each copula diagram represent the best-fitted bivariate copula where "C" is Clayton copula, "F" is Frank copula, "G" is Gumbel copula, "N" is Gaussian copula, and "t" is Student's t copula. The letter "T" indicates the tree number for each row. The variables joint in the bivariate copulas are indicated as follows: 1 for saturated hydraulic conductivity; 2 for scaling parameter α; 3 for shape parameter n; 4 for saturated water content; 5 for residual water content; 6 for unit weight; 7 for effective cohesion; and 8 for effective friction angle.

4.2. Regression model based on Multilayer Perceptron

The topology of the neural network model constructed in this study is explained in Figure 5. The initial hidden node number is assumed as five. The regression model performance is assessed by applying a higher number of hidden nodes, yet the increase in accuracy is negligible. Thus, the selection of five number of hidden nodes is justified in this study. The limit state function equation can then be derived. The R^2 coefficients for the training dataset range from 0.9882 to 0.9898. The MAE and RMSE for the training dataset vary from 0.0271 to 0.0366 and 0.0311 to 0.0376, respectively. The testing dataset show R^2 coefficients between 0.9477 and 0.9557. It also displays MAE and RMSE from 0.0641 to 0.9542 and 0.0772 to 0.1227, respectively. As the value of R^2 coefficients are mostly approaching one, while the values of MAE and RMSE are close to zero for both training and testing dataset, the performance of the regression model is acceptable.

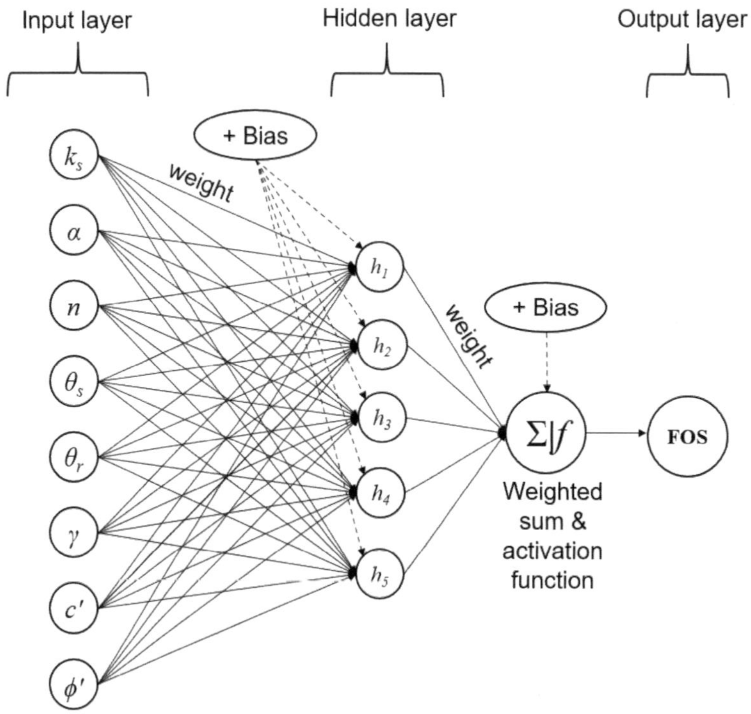

Figure 5. Topology of multilayer perceptron regressor in this study.

4.3. Threshold for landslide initiation

In this study, the probability of failure ranges from 0.0001 from 0.051 for different antecedent rainfall condition (Figure 6). The mean factor of safety display values between 0.949 and 1.782. The probability of failure is generally increasing, while the mean factor of safety is decreasing with the increase in the rainfall duration and cumulative precipitation. This can be explained that the shear strength of the soil is reduced due to seepage in the slope. During the initial stage, prior to application of antecedent rainfall, the probability of failure is 0.0001, which can be categorised as "good" performance level. When the rainfall was applied for one day, the probability of failure increase to 0.00094. At this stage, the performance level is approaching the threshold of "above average". Further application of antecedent rainfall from day two to day seven has caused a rise in the probability of failure from 0.0014 to 0.0056, with a reduction in the mean factor of safety from 1.602 to 1.196. During this 6-day rainfall duration, the slope shows the performance level of "above average". On day two and day five, no increase in rainfall infiltration has been observed. However, the probability of failure is kept increasing due to redistribution of rainwater in the slope. The performance level of the slope decline to "below average" upon application of 9-day antecedent rainfall. The slope is considered fail when the performance level reaches the "poor" threshold. The corresponding probability of failure is 0.0051. Therefore, the landslide can be defined as the condition when the "poor" performance level is reached or exceeded. The 9-day antecedent rainfall with a cumulative amount of 273 mm depth, combined with a major storm of 74 mm depth is therefore considered as the rainfall threshold which may trigger landslide in the study area.

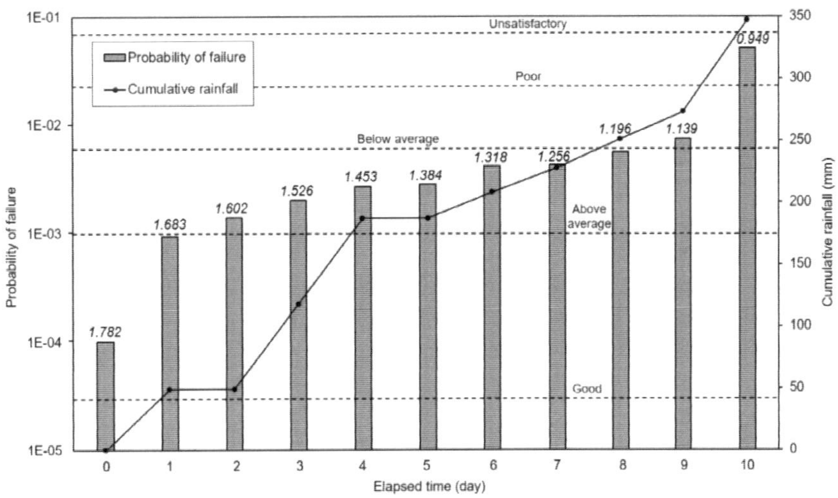

Figure 6. Probability of failure for different duration of antecedent rainfall. The values above each histogram indicate the mean factor of safety.

5. Conclusion

This study has presented a relatively computationally, inexpensive reliability-based approach to evaluate the physical threshold of landslide due to antecedent rainfall. The application of the framework has been demonstrated by taking a case study from Sabah, Malaysia. The uncertainties considered are the eight dependent random variables (saturated hydraulic conductivity, scaling parameter α and shape parameter n of SWCC, saturated water content, residual water content, unit weight, effective cohesion, effective friction angle), which were efficiently sampled using the Latin Hypercube Sampling. The dependencies of the multivariate were also constructed using the flexible D-vine copula, in which the best-fitted bivariate copula can be selected as the building block. The Multilayer Perceptron network, which can capture the non-linear relationship between input and output variables is adopted to derive the limit state function from the coupled seepage-deformation and shear strength reduction analysis. By taking the limit state function as the surrogate model for the expensive deterministic analysis, the Monte Carlo Simulation is realised to compute the probability of failure and mean factor of safety. It is found that the proposed model displays an acceptable performance with the R^2 coefficients approaching one, while the MAE and RMSE are close to zero for both training and testing dataset. The expected performance level can then be assessed based on the probability of failure, which provides a more representative index for the slope stability. It has been demonstrated in this study that the landslide may occur when the "poor" performance level is reached or exceeded. The physical threshold in the study which may trigger landslide pertains to a cumulative rainfall depth of 273 mm for 9-day antecedent rainfall and a major storm of 74 mm depth.

Acknowledgement

The work described was supported by the grant of UTM Fundamental Research 20H95. The authors acknowledge the assistance from the Public Works Department of Sabah and the Department of Meteorology, Malaysia, for providing the soil and rainfall data, respectively.

References

[1] Guzzetti F, Peruccacci S, Rossi M and Stark C P 2007 *Meteorol. Atmos. Phys.* **98** 3–4
[2] Rahardjo H, Li X W, Toll D G and Leong E C 2001 *Geotech. Geol. Eng.* **19**
[3] Wei J, Heng Y S, Chow W C and Chong M K 1991 *Ninth Asian Regional Conference on Soil Mechanics and Foundation Engineering*
[4] Dou H, Han T, Gong X and Zhang J 2014 *Eng. Geol.* **183**
[5] Ering P and Babu G L S 2016 *Eng. Geol.* **208**
[6] Davidovic N, Prolovic V and Stojic D 2010 *Facta Univ. Ser. Archit. Civ. Eng.* **8** 1
[7] Mahmood K, Kim J M, Khan, Safdar M and Khan U 2018 *KSCE J. Civ. Eng.* **22** 5
[8] Van Genuchten M T 1980 *J. Soil Sci. Soc. Am.* **44** 5
[9] Mualem Y 1976 *Water Resour. Res.* **12** 3
[10] Silvestrini R T, Montgomery D C and Jones B 2013 *Qual. Eng.* **25** 2
[11] Sheela K G and Deepa S N 2013 *Math. Probl. Eng.* **2013**
[12] Kang F, Li J and Xu Q 2017 *Int. J. Numer. Anal. Methods Geomech.* **41** 18
[13] US Army Corps of Engineers 1997 *Technical Letter* 1110-2-547

IOP Conf. Series: Materials Science and Engineering 1153 (2021) 012003 doi:10.1088/1757-899X/1153/1/012003

Numerical Modelling of Pollutant Transport in a Straight Narrow Channel using Upwind Finite Difference Method

B F Yip[1], N A F Alias[1] and E H Kasiman[2]

[1]School of Civil Engineering, Universiti Teknologi Malaysia, Malaysia
[2]Department of Water and Environmental Engineering, School of Civil Engineering, Universiti Teknologi Malaysia, Malaysia
E-mail: yipbaofang@gmail.com

Abstract. River pollution has become an increasingly severe issue, resulting in the rapid depletion of freshwater resources. An accurate prediction of the transport of the pollutants is vital for effective mitigation measures. This paper presents the numerical modelling of pollutant transport in a straight narrow channel. The pollutant transport is modelled using an advection-diffusion equation. The equation is discretized using Finite Difference Method (FDM) with an explicit upwind scheme for the temporal term. The resulting system of simultaneous equations is solved using a direct method. The numerical model is validated against an existing analytical solution, and the results show good agreement between numerical prediction and analytical solution. The validated numerical model is applied to different cases of pollutant release mechanisms involving continuous and instantaneous pollutant releases. In both cases, the numerical model can capture the physics of the problem and able to provide valuable information on the time and spatial evolution of pollutant concentration. The model is extendable to higher dimension owing to its simplicity of discretization.

1. Introduction

Over the past few decades, we have been facing serious issue with river [1]. This issue is critical as river is the one of the main water resources for our daily activities. River pollutant not only affect the environment but human wellbeing as well. If this problem is not resolved quickly, it is expected that we will be facing a serious depletion in clean water resources in near future. In order to mitigate this problem, we need a tool that can be used to predict the transport of pollutant effectively.

Experiments, measurements and observations have been widely used to investigate the physical process of pollutant transport [2]. However, the drawbacks with experimental investigation are cost and control. The cost of running even a simple experiment is not cheap. The total cost includes the cost of material, construction and manpower. Additionally, in an experimental setup, it is difficult to control or separate all the unnecessary variables and time-consuming. Extra effort is needed to filter the undesirable effect from the experimental result. In most cases, some of this uncertainty cannot be eliminated and remains in the result. Moreover, the experiment provides limited information and the results may be subjected to human and measurement errors.

With the advancement of computer and technology, numerical field has been developed rapidly and a remarkable progress has been done in the development of numerical methods such as finite difference, finite element and finite volume [2]. This allows the application of mathematical or partial differential equations that describe the physics of natural phenomenon on every flow problem. In the context of pollutant transport problem, numerical simulation has been used to predict the behaviour of pollutant pattern under specific environmental conditions [3,4]. From the perspective of numerical method, the transport of pollutant can be modelled using advection-diffusion equation (ADE) [5]. Advection is the transport of a property due to the background fluid velocity. In the case of pollutant, this property is the pollutant concentration. Diffusion on the other hand, is the net movement of this concentration from high concentration to low concentration. ADE has been used to predict the diffusion of pollutant concentration as it is being carried away by the velocity of the water.

The pollutant transport phenomenon is extremely complex. Traditionally, the prediction of such transport quantities was limited to the laboratory studies and simplified theories. This former is used to model the realistic case but is expensive and time consuming. The latter on the other hand is simple and easy to use but limited to simple problem and in most cases does not provide a real representation of the problem. With the advancement of computer technology and numerical method, a numerical model is developed to enhance the understanding of the physical process that governs the pollutant transport phenomenon. This numerical model serves as an alternative tool to both experimental and simplified theories. A numerical method provides a much more realistic representation of the problem at a fraction of the cost of running a full experimental work.

The aim of this study is to solve a one-dimensional pollutant transport model using ADE. The objectives are to discretize a one-dimensional ADE using an explicit upwind FDM, to validate the numerical model against existing analytical solution and to determine the behaviour of pollutant concentration for different pollutant release mechanisms. In this study, the numerical modelling of pollutant transport is undertaken. The ADE is used as the governing equation of the problem. The ADE is discretized using an explicit upwind finite difference method. The numerical solution is validated with existing analytical solution before being used to simulate few cases of pollutant transports problems in a straight narrow channel.

2. Literature Review

In the context of pollutant transport, there are four mechanisms that need to be accounted. These are advection, shear flow dispersion, turbulent diffusion, and molecular (Fickian) diffusion. If the flow is laminar, the effect of shear and turbulence is not significant. In this case, only the advection and diffusion play an active role. These mechanisms can be modelling using ADE. ADE describe combined effect of advection and diffusion in fluid medium. Separately, advection is the transport mechanism of a substance or conserved property by a fluid due to the fluid's bulk motion while diffusion is the net movement of particles from high concentration to low concentration. The theoretical behaviour of both advection and diffusion is presented in Figure 1(a) and 1(b) respectively.

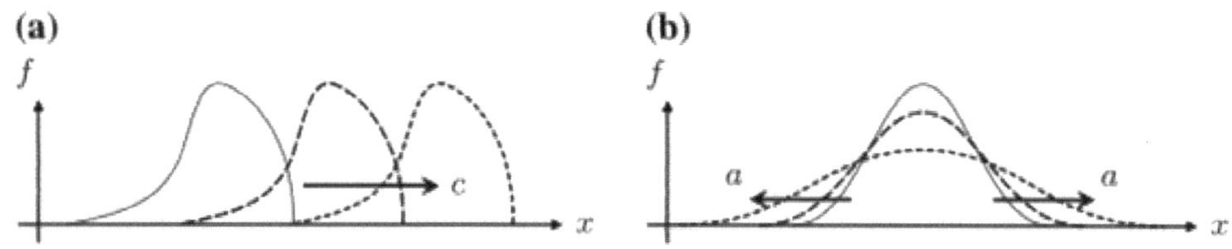

Figure 1. The behaviour of (a) advection and (b) diffusion transport mechanism.

The simplest and most widely used numerical method to solve directly differential equations is the FDM. The solution of differential equation using FDM is called a strong form solution. The strong form stem from the fact that the FDM discretized the differential equation without reducing the order of the differential equation like most recent numerical techniques, for example Finite Element Method (FEM), Finite Volume Method (FVM) and Boundary Element Method (BEM), do. Most people use FDM due to its simple discretization and implementation [6]. Higher order solution can be obtained by simply by changing the order of discretization and does not alter much the subsequence step of the overall algorithm. The implementation of both the initial and boundary condition in FDM is also straightforward intuitive. Both the Dirichlet (variable) and Neumann (derivative of variable) boundary condition can be implemented using a simple substitution in the discretized matrix. Although FDM are usually used for problems involving simple domain, the method is undoubtedly able to provide valuable information that cannot be obtained from analytical solution alone. This is especially true when the boundary condition is complex and not smooth. Taking these advantages into consideration, the FDM is used for the modelling of pollutant transport in this study.

3. Mathematical Model

3.1. *Governing equation*

The general equation for one-dimensional solute transport considering advection-diffusion without source or sink term is as follow:

$$\frac{\partial c}{\partial \mathrm{t}} = D\frac{\partial^2 c}{\partial x^2} - u\frac{\partial c}{\partial x} \tag{1}$$

where, c is the concentration of solute, *u* is the velocity and *D* is the diffusion coefficient. This is so called advection-diffusion equation (ADE). The first term on the right-hand side of Equation (1) is the diffusion term while the second one is called the advection term. Due to unsteady nature of the equation, both boundary and initial conditions are required to properly defined the problem.

Finite difference approximations can be used to represent the derivatives in ADE. The derivation of ADE will be discussed in the following section.

3.2. *Finite Difference Methods*

3.2.1. Discretization of advection term. Advection can be defined as the process of transporting mass which being carried by the flow from one point to another. Equation (2) is the pure advection term in the ADE. $\partial c/\partial x$ is represented by backward finite difference approximation. *u* is assumed as a constant which represent the flow velocity in the channel.

$$-u\frac{\partial c}{\partial x} = -u\frac{c_i - c_{i-1}}{\Delta x} \tag{2}$$

3.2.2. Discretization of diffusion term. Diffusion is the mixing of dissolved chemical due to the random motion of molecules within the fluid. The solute moves from high concentration region to low concentration region based on Fick's first law of diffusion.

$$D\frac{\partial^2 c}{\partial x^2} = D\frac{c_{i+1} - 2c_i + c_{i-1}}{\Delta x^2} \tag{3}$$

Equation (3) is the diffusion term in the ADE. Equation (3) is represented by finite difference equation using the second-order central differencing of order $(\Delta x)^2$. D is assumed as a constant which represent diffusion coefficient of the pollutant transport in medium such as liquid and air. In Equation (2) and (3) above, the subscript i represent the spatial index of the variables. The summation and subtraction of i means that the variables are location either to the right or the left of the main variables, respectively.

3.2.3. Discretization of temporal term. There are two methods for the discretization of temporal term in finite difference such as explicit and implicit method. The implicit method assumes that all or most of the spatial variables are unknowns that need to be solved at each time steps. This requires that all the variables to solve simultaneously, resulting in the creation of a system of simultaneous equation. In any numerical method, the solution to the simultaneous equation is the most expensive part of the whole solution. Explicit method on the other hand assumed that all spatial variables are known and only one variable need to be solved at any given time step. Therefore, it simpler and requires less computational effort as compared to implicit method. In this study, the explicit method is used. The general form of the explicit method is given as

$$\frac{\partial c}{\partial t} = f\ (c) \tag{4}$$

$$\frac{c_i^{n+1} - c_i^n}{\Delta t} = f(c)^n \tag{5}$$

where the subscript n and $n + 1$ represent the previous and current time step, respectively.

3.2.4. Full discretization of ADE. Equation (1) can be fully discretized by substituting Equation (2), (3) and (5) resulting in the following discretized equation.

$$\frac{c_i^{n+1} - c_i^n}{\Delta t} = D\frac{c_{i+1}^n - 2c_i^n + c_{i-1}^n}{\Delta x^2} - u\frac{c_i^n - c_{i-1}^n}{\Delta x} \tag{6}$$

Equation (6) can be solved by rearranging all the terms with superscript n to the right-hand side and solving for the term $n + 1$ only. Although the explicit method is easy to solve, it suffers from stability issue. The method is only stable if the following criteria is fulfilled.

$$D\frac{\Delta t}{(\Delta x)^2} \leq \frac{1}{2} \tag{7}$$

$$u\frac{\Delta t}{\Delta x} \leq 1 \tag{8}$$

Equation (8) and (9) are called the diffusion and Courant numbers, respectively.

3.3. Numerical simulation

3.3.1. Initial and boundary conditions. In order for the problem to be well defined, initial and boundary conditions is required. The initial condition is related to the temporal term in the ADE, represented by the superscript n. Because the temporal term is only first derivatives, only one initial condition is required. This is given by:

$$C\ (0, x) = c(x) \tag{9}$$

where $C(0, x)$ represent all the spatial variables at time step, $t = 0$. The boundary condition on the other hands is related to the spatial variables and because the highest derivatives in the ADE is second order, the boundary condition needed to be fully defined the problem is two. In the case where the inlet is prescribed with known value and the pollutant is allowed to exit at the outlet boundary, the boundary conditions are given as

$$C\,(t, 0) = c(t) \tag{10}$$

$$\frac{\partial c}{\partial t}(t, L) = 0 \tag{11}$$

Equation (10) and (11) are called Dirichlet and Neumann boundary conditions, respectively.

4. Numerical Experiments and Discussion

This study assumed that the pollutant is transported downstream in a straight narrow channel. The unknown quantity is the concentration. The concentration is assumed to varies in just one direction. Therefore, the problem is assumed to be one-dimensional. As the concentration varies at each time step, the problem is considered as unsteady. The transport of the pollutant is modelled using an ADE. A simple explicit upwind FDM is used to solve the advection diffusion equation. The numerical solution of ADE provides a prediction of pollutant concentration distributions along the straight channel.

4.1. Model validation

The discretized ADE using FDM is given by Equation (7). In order to ensure that both the derivation and code is correct, validation need to be undertaken. For the ADE, an analytical solution exists. The analytical solution assumed that the initial condition of the concentration is given by the following equation.

$$C(0, x) = e^{\frac{-(x-x_0)^2}{D}} \tag{12}$$

where x_0 is the location of the peak concentration.

The diffusion coefficient, D, is assumed to be $0.005\text{m}^2/\text{min}$ while the background velocity, u, is given by $0.05\text{m}/\text{min}$. The length of the domain is assumed to be 2.5m. The domain is discretised into 313 nodes with a spacing of 0.008m. The total duration of the simulation is 50min and the time step used is 0.001min. In this case, the diffusion and Courant numbers are given by 0.0781 and 0.0063, respectively. As both values are lower than their critical limit, the solution is stable.

Figure 2 shows the concentration at various time steps between numerical results and analytical solution. At time $t = 0$min, the pollutant concentration is given by a symmetric bell shape. This concentration is given by Equation (12). The concentration has a maximum value of 1 kg/m^3, located at $x = 0.5$m. As time progresses, the concentration drops rapidly at first due to diffusion and advected to the right due to the background velocity. At $t = 10$min, the maximum concentration value reduces to $0.2\text{kg}/\text{m}^3$ and occur at $x = 1$m. At the next time step, $t = 20$min, the reduction in the maximum concentration value is very subtle. However, the maximum concentration value has been advected further to the right at $x = 1.6$m.

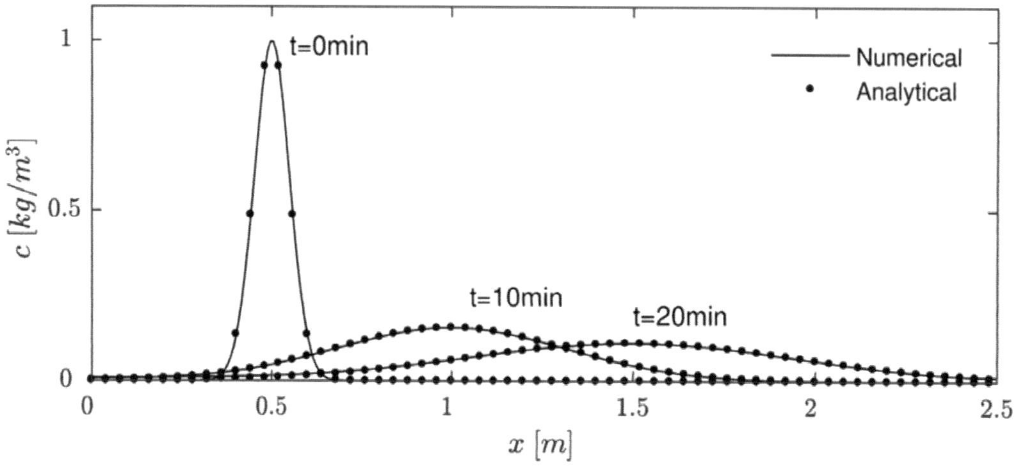

Figure 2. Concentration at various time steps between numerical result and analytical solution.

At each time steps, it can be seen that the agreement between numerical prediction and analytical solution is excellent. This shows that the FDM discretization of the ADE is correct and the numerical code is free from error.

4.2. Simulation 1: Continuous Pollutant Release

Having validated the numerical model against existing analytical solution, this section seeks to provide the simulation of pollutant transport for a continuous pollutant release. It is important to find out the behaviour of the pollutant concentration at various part of the domain if say, a pollutant is release continuously. Before the simulation is undertaken, it is important that a correct initial and boundary conditions is imposed so that the correct physics is simulated.

Initially, it is assumed that the river or channel is not polluted yet. In this case, the pollutant concentration over the entire domain is zero. The initial condition that needs to be imposed is given as:

$$C\,(0, x) = 0 \tag{13}$$

As we need to impose continuous pollutant release at the inlet, the boundary condition at the inlet needs to be a function of time. Initially, the value is zero. As the pollutant is released at the inlet, the pollutant concentration increases. After some time, the concentration reaches maximum value and plateaued. The boundary condition at the inlet is given as follows:

$$t = 0, \qquad c(x) = 0$$

$$0 < t < 10, \qquad c(x) = \frac{1}{10}t$$

$$t > 10\text{t}, \qquad c(x) = 1$$

At the outlet, the pollutant concentration is allowed to exit the boundary. Therefore, Neumann boundary condition is applied as given by Equation (11).

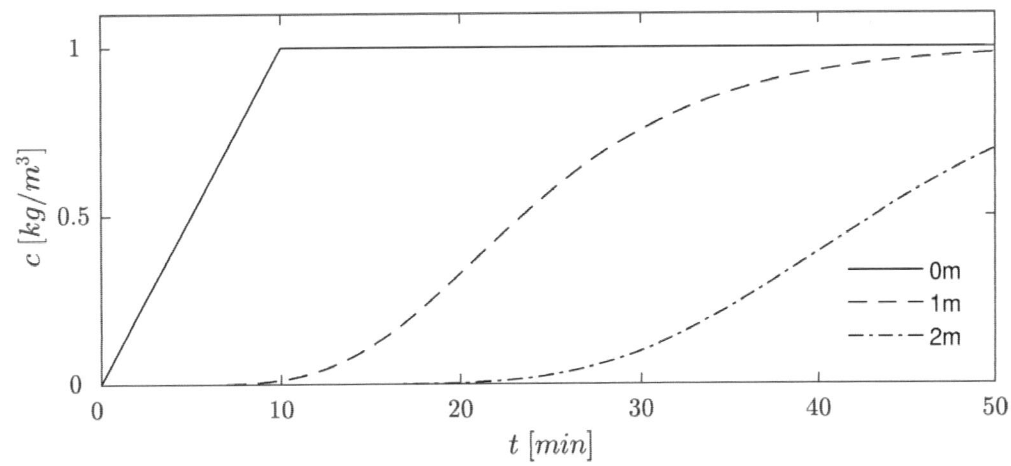

Figure 3. Concentration time-histories at various location for continuous pollutant release

The concentration of pollution at $x = 0\text{m}$, $x = 1\text{m}$ and $x = 2\text{m}$ are given in Figure 3. The inlet is located at $x = 0\text{m}$. From the figure, it is clear that the correct boundary condition is imposed at the inlet. At $t = 0\text{min}$, the concentration at all these locations are zero, as time progresses, the concentration starts to increase. Because the location differs, the time required for the concentration to initiate is slightly delayed. At $x = 1\text{m}$, the concentration starts to increase at around $t = 10\text{min}$ while for $x = 2\text{m}$, this occurs at around $t = 20\text{min}$.

Another interesting observation that can be made from Figure 3 is the shape of the concentration profile. At the inlet, the pollutant concentration has a distinct profile, a sharp change in profile at $t = 10\text{min}$. This profile changes completely for second and third location; the profile is much smoother than the inlet profile. These smooth changes are caused by the diffusion term in the ADE. As the advection term is only responsible for the movement of the concentration, the only plausible explanation for the changes in the profile is due to the diffusion.

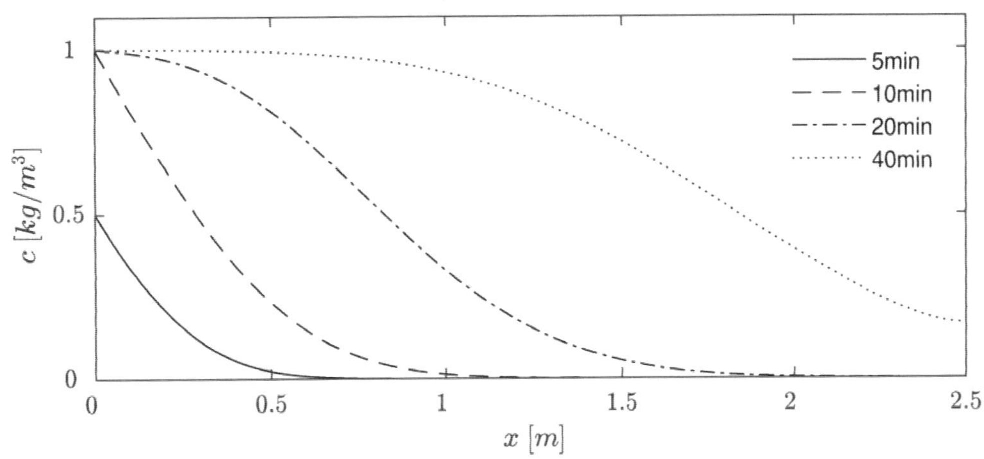

Figure 4. Concentration over the whole domain at various time steps for continuous pollutant release

The spatial variation of the concentration over the whole domain at different time steps is given in Figure 4. From the figure, we can clearly see the evolution of the concentration over the whole domain. Initially, the concentration is zero over the whole domain, as time progresses, the concentration at the inlet increases due to continuous release of pollutant. At the same time, due the background velocity, the concentration is being advected to the right. As explained earlier, the diffusion term in the ADE has a smoothing effect and forces the profile varies smoothly over the whole domain. With further increase in time, the pollutant concentration starts to reach peak the maximum value. If the simulation is continued further, the value of concentration will reach a maximum value of 1kg/m^3 over the whole domain.

4.3. Simulation 2: Pollutant Release Over a Short Period

The exact solution for pollutant release over a short period is

$$
\begin{aligned}
&t = 0, && c(x) = 0 \\
&0 < t < 10, && c(x) = \frac{1}{10}t \\
&10 < t < 15, && c(x) = 1 \\
&15 < t < 25, && c(x) = \frac{1}{10}t + 2.5 \\
&t > 25, && c(x) = 0
\end{aligned}
$$

In this section, the test case also divided into two types; concentration time-histories at various location for pollutant release over a short period and concentration over the whole domain at various time steps for pollutant release over a short period. In this case, the pollutant is released in a limited time period.

In concentration time-histories at various location for continuous pollutant release, the time is set as $0 \leq t \leq 50\text{s}$. In concentration over the whole domain at various time steps for continuous pollutant release, the value of the distance is set as $0 \leq L \leq 2\text{m}$. The velocity in all direction is assumed constant. The dispersion pattern and behaviour of the pollutant concentration are observed in Figure 5 and 6.

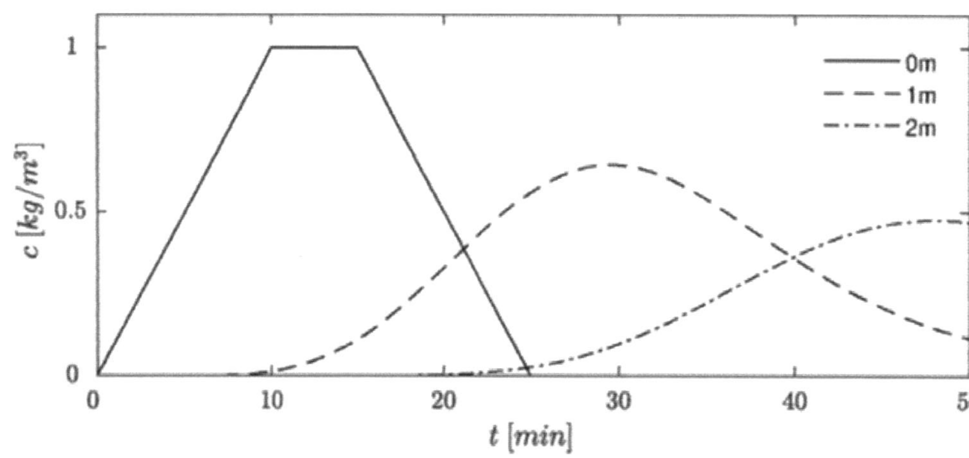

Figure 5. Concentration time-histories at various location for pollutant release over a short period

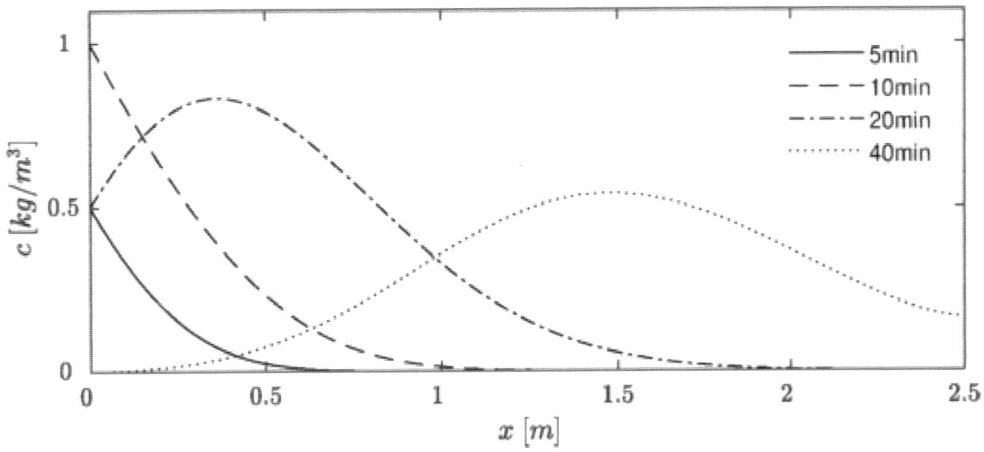

Figure 6. Concentration over the whole domain at various time steps for pollutant release over a short period

5. Conclusion

In this paper, a one-dimensional transport of pollutants is solved using finite difference method (FDM). The pollutant transport dependent on few key parameters such as diffusion coefficient and the background velocity. The results obtained using FDM shows good agreement when compared to analytical solution. Numerical simulations of pollutant transport for continuous and fixed period release shows that the numerical model is capable of simulating the physics of the problems correctly.

Acknowledgement
The authors would like to acknowledge the financial supports by Ministry of Higher Education Malaysia and Universiti Teknologi Malaysia through Skim Latihan Akademik IPTA (SLAI) scholarship.

References

[1] Yap C K, Peng S H T and Leow C S 2019 Contamination in Pasir Gudang Area, Peninsular Malaysia: What can we learn from Kim Kim River chemical waste contamination? *J. Humanit. Educ. Dev. JHED* **1** 84–7

[2] Kuzmin D 2010 *A Guide to Numerical Methods for Transport Equations* (Germany: Friedrich-Alexander-Universität Erlangen-Nürnberg)

[3] Kachiashvili K, Gordeziani D, Lazarov R and Melikdzhanian D 2007 Modeling and simulation of pollutants transport in rivers *Appl. Math. Model.* **31** 1371–96

[4] Ani E-C, Wallis S, Kraslawski A and Agachi P S 2009 Development, calibration and evaluation of two mathematical models for pollutant transport in a small river *Environ. Model. Softw.* **24** 1139–52

[5] Yadav N, Yadav A and Kim J H 2016 Numerical solution of unsteady advection dispersion equation arising in contaminant transport through porous media using neural networks *Comput. Math. Appl.* **72** 1021–30

[6] Langtangen H P and Linge S 2017 *Finite Difference Computing with PDEs: A Modern Software Approach* (Springer International Publishing)

[7] Ahmed S G 2012 A Numerical Algorithm for Solving Advection-Diffusion Equation with Constant and Variable Coefficients *Open Numer. Methods J.* **4** 1–7

[8] Anderson J D 1995 *Computational Fluid Dynamics: The Basics with Applications* (McGraw-Hill)

[9] Danaila I, Joly P, Kaber S M and Postel M 2007 *An Introduction to Scientific Computing: Twelve Computational Projects Solved with MATLAB* (Springer Science & Business Media)

[10] Dehghan M 2004 Numerical solution of the three-dimensional advection–diffusion equation *Appl. Math. Comput.* **150** 5–19
[11] Hoffmann K A and Chiang S T 2000 *Computational Fluid Dynamics: Volume 1* vol 1 (Wichita, Kansas, USA: Engineering Education System)
[12] Kajishima T and Taira K 2017 Finite-Difference Discretization of the Advection-Diffusion Equation *Computational Fluid Dynamics: Incompressible Turbulent Flows* ed T Kajishima and K Taira (Cham: Springer) pp 23–72
[13] Shaari K Z K and Awang M 2015 *Engineering Applications of Computational Fluid Dynamics* (Cham: Springer)
[14] Kumar A, Jaiswal D K and Kumar N 2010 Analytical solutions to one-dimensional advection–diffusion equation with variable coefficients in semi-infinite media *J. Hydrol.* **380** 330–7
[15] Kumar A, Jaiswal D K and Kumar N 2009 Analytical solutions of one-dimensional advection-diffusion equation with variable coefficients in a finite domain *J. Earth Syst. Sci.* **118** 539–49
[16] Pochai N and Deepana R 2011 A Numerical Computation of Water Quality Measurement in a Uniform Channel Using a Finite Difference Method *Procedia Eng.* **8** 85–8
[17] Recktenwald G W 2004 Finite-Difference Approximations to the Heat Equation
[18] Sankaranarayanan S, Shankar N J and Cheong H F 1998 Three-dimensional finite difference model for transport of conservative pollutants *Ocean Eng.* **25** 425–42
[19] Savović S and Djordjevich A 2012 Finite difference solution of the one-dimensional advection–diffusion equation with variable coefficients in semi-infinite media *Int. J. Heat Mass Transf.* **55** 4291–4
[20] Szymkiewicz R 2010 Numerical Solution of the Advection-Diffusion Equation *Numerical Modeling in Open Channel Hydraulics* Water Science and Technology Library ed R Szymkiewicz (Dordrecht: Springer Netherlands) pp 263–300

IOP Conf. Series: Materials Science and Engineering 1153 (2021) 012004 doi:10.1088/1757-899X/1153/1/012004

Influence of Dam to Rainfall-Runoff Response in a Tropical Climate – A Case Study of Selangor River Basin, Malaysia

A F Bahar[1], **Z Yusop**[1,2], **N E Alias**[1,2], and **M W A Ramli**[1]
[1]School of Civil Engineering, Faculty of Engineering, Universiti Teknologi Malaysia, 81310 Johor Bahru, Johor, Malaysia.
[2]Centre for Environmental Sustainability and Water Security (IPASA), Universiti Teknologi Malaysia, 81310 Johor Bahru, Johor, Malaysia.
Email: fitrybahar61@gmail.com

Abstract. This paper discusses the rainfall-runoff response by comparing the runoff coefficient, *c* between two catchments of different characteristics within the Selangor River Basin, namely Catchment A (CA) and Catchment B (CB). Catchment B is a sub-catchment of Catchment A, whereby CA is about 6.8 times larger than to CB. The landscape of CB is largely covered by forest, has higher river slope and a dam completed in year 2004 at upstream of the basin. Annual mean-areal-rainfall (MAR) and total runoff from 1990 until 2018 was used for the rainfall-runoff coefficient, *c* calculations. The rainfall-runoff response was assessed by separating the analysis into 2-time frames; 1990 to 2000 (before dam construction) and 2007 to 2018 (after dam construction). The analysis before dam construction showed that CB has higher runoff response compared to CA. This was due to the difference in land use, catchment slope and lesser water abstractions compared to CA. The runoff coefficient, *c* for CA was in the range of 0.19-0.48, while CB was in the range of 0.59-0.79. This means that 19% to 48% of the rainfall was converted into runoff for CA, and 59% to 79% of the precipitation turned into runoff for CB. As suspected, after the dam construction the *c* coefficient for CB changed to 0.28-0.89, indicating a higher variability in the catchment response due to dam operations and control at the upstream.

1. Introduction

Flooding is a natural environmental event that occur around the countries in the world. Land use and land cover change (LULC) is the main factor that influences the hydrological cycle and the availability of water resources. The change in LULC alters the interception, infiltration rate and evapotranspiration (EP) [1]. Runoff can be defined as the portion of precipitation that makes its way towards rivers or oceans as surface or subsurface flow. In general, runoff is generated through two different mechanism which are saturation excess and infiltration excess. Saturation excess is generated when the soil become fully saturated with water then additional precipitation cannot infiltrate [3]. Infiltration excess occurs when the rainfall intensity becomes higher than the infiltration rate, and becomes runoff flowing over land to a different location [4].

The runoff coefficient is widely used and is an important parameter describing the basin response either in an annual or event basis. Annual runoff coefficient is the total runoff over total precipitation in a year [5] while event-based runoff coefficient are the ratio of total flow over total rainfall for a specific storm event. The use of event flow instead of total annual flow allows to investigate rainfall-runoff response for a single event. However, by using total flow, the response of single event is combined with pre-event flow condition. Meaning, by event based, high flow in the wet season or low flow during summer may be observed [5]. According to Thompson (2006), runoff coefficient, *c*, can be defined as the dimensionless ratio intended to indicate the amount of runoff generated by the watershed to the amount of precipitation received. In general, a larger value of *c* to the area with low infiltration and high runoff such as pavement, steep gradient while a lower value of *c* to area with high infiltration and low runoff such as forest and flat land.

Land use changes are mostly induced by human activities, for example, deforestation, afforestation, and farmland reclamation have led to significant agricultural-land-pattern changes and eventually affect the hydrological processes such as evapotranspiration (ET), interception, infiltration. This will finally change the surface and subsurface flows [7]. Land use is the utilization of land resources by humans for various land activities and purposes such as for transport (road and highway), agriculture (farm), residential (housing and settlement), commercial (business and factories) and power (dam and reservoir). Dams and reservoirs impact the environment through their presence in the landscape by altering basin connectivity and modifying thermal, hydrologic and sediment regimes [8]. The water in a river is a complex water system characterized by numerous river reaches and reservoir due to the influences of dams [9]. Meanwhile, another effect of dams regulation is reducing peak flow during the flooding season and releasing during the dry season due to change in the river flow regime [10].

The Selangor River Basin (SRB) (**Figure 2**), was selected for this study because it is the major source of water supply for Selangor and Federal Territory. It is also prone to rapid land use change and development due to its proximity to Kuala Lumpur and other major townships. As in the parts of Peninsular Malaysia, the climate is characterized by all year round warm and humid with mean daily temperature from 25°C to 32°C [12]. According to Wong et al. (2016) rainfall is characterized by two rainy seasons, which are southwest monsoon (SWM) from May to September and the northeast monsoon (NEM) from November to March. The SRB receives an annual rainfall average of 2154 mm/year. The rainfall during the northeast monsoon is higher (921 mm) than during the southwest monsoon (765 mm). The monthly rainfall shows a bimodal distribution with peak in April and November which coincide with inter-monsoon periods [14].

The objectives for this study are:

1. To calculate the rainfall runoff response and rainfall characteristics on different landscapes, geology and size of catchment.
2. To study the effect of dam on river flow regime of the SRB.

2. Study Area

Figure 1 shows river basins in Selangor and location of dams. The Selangor River Basin (SRB) is located in the state of Selangor and within the district of Hulu Selangor, Gombak and Kuala Selangor. **Figure 2** shows the SRB and the locations of rainfall and water level stations. The basin has a total area of 1820km^2 and has been undergoing rapid land development and land use change since the 1970s [15]. There are two river flow stations in the basin located at Rantau Panjang and Ampang Pecah. Two water supply dams were constructed in the basin as shown in **Figure 1**. Sungai Selangor Dam was completed and operational in year 2004 and Sungai Tinggi Dam was completed in 1997 (**Table 1**). The Sungai Selangor Dam has a catchment area of 197 km^2, located in the upper catchment of SRB with supply capacity of 1,100 MLD or 12.7 m^3/s. The Sungai Tinggi Dam is smaller (40 km^2), located in the upper reaches of Sungai Buloh, and it was designed to regulate the flow at the existing Batang Berjuntai water supply intake. The Sungai Selangor and Sungai Tinggi Dams were designed to regulate the river flow to give a total net system yield of 3016.5 MLD [16]. The SRB contains eight (8) water treatment plants (water intake), as shown in **Table 2** and **Figure 3**. In this analysis, the SRB is divided into two catchments namely Catchment A (CA) and Catchment B (CB). Catchment A is the whole basin of SRB while CB is the sub-basin of CA and situated at its upstream.

IOP Conf. Series: Materials Science and Engineering 1153 (2021) 012004 doi:10.1088/1757-899X/1153/1/012004

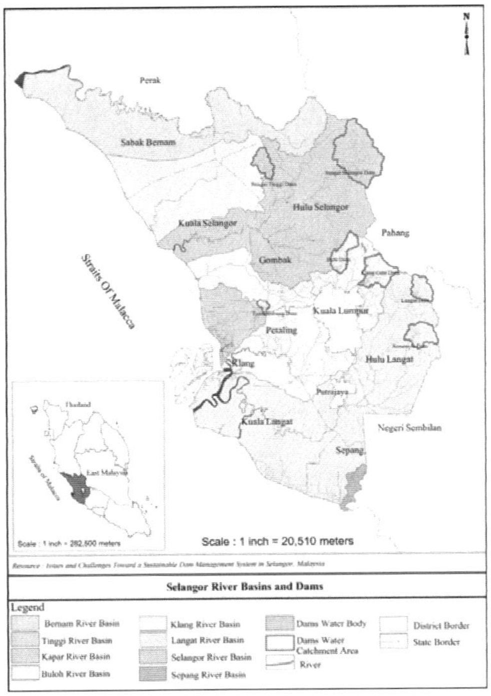

Figure 1. River basins and dams in Selangor.

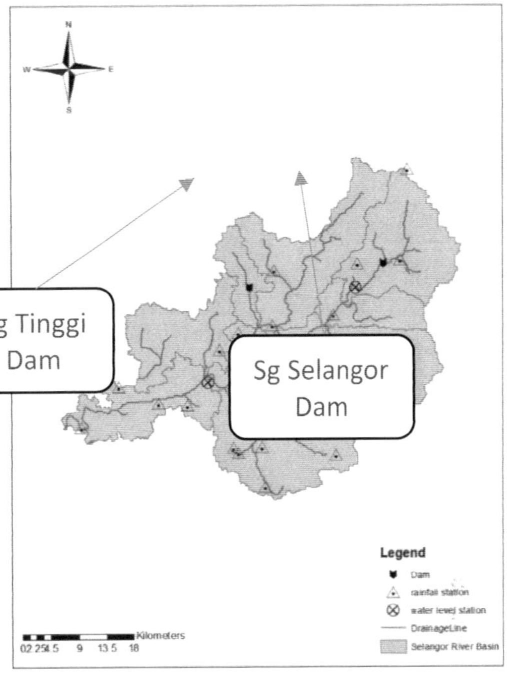

Figure 2. Selangor River Basin (SRB).

2.1. Catchment A

Catchment A (CA) has an area of 1502 km^2 and contains seven sub-basins (**Figure 4 (a)**). About 38% of the catchment is still forested, mostly in the upper catchment. Rubber plantation is the second largest land use (21.6%) followed by residential area which include institution, village, industrial and urban (20.7%), oil palm tree (9.8%), ex-mining land (6.2%) and mangrove (3.7%) as shown in **Figure 4 (b)**. The urban and residential areas are mostly located in the downstream. There are two water supply reservoirs in CA, namely Sungai Selangor Dam in the upstream of Sungai Selangor and Sungai Tinggi Dam in the upper reaches of Sungai Buloh. **Figure 4 (d)** shows a drainage map where CA have seven (7) number of stream order, u with 275km of perimeter, P, stream length, Lu (3875), mean stream length, Lsm (553.51), basin length, Lb (83.61km), drainage density, Dd (2.58km/km^2), stream frequency, Fs (7.15), drainage texture, Rt (39.05), form factor, Rf (0.215), circulatory ratio, Rc (0.250), elongation ratio, Re (0.523), compactness constant, Cc (2.00), length of overland flow, Lg (0.78) and shape index, Sw (4.65) as listed in **Table 3. Figure 4 (c)** shows catchment elevation ranging from 33m to 260m and the steepest slopes are mostly found in the upstream.

2.2. Catchment B

Catchment B (CB) is about 220 km^2 (**Figure 5 (a)**), and only about 79.3% of the land use is still forested. Sungai Selangor Dam is located within the catchment. Residential area which included urban and villages comprises about 9.8%, rubber plantation (5.4%), palm oil tree (4.4%) and ex-mining land (1.1%) (**Figure 5 (b)**). Meanwhile, 84.7% of the catchment area is considered impervious and may be a major source of runoff. **Figure 5 (d)** shows a drainage map whereby CB has six (6) number of stream order, u with 97km of perimeter, P, stream length, Lu (528), mean stream length, Lsm (87.96), basin length, Lb (28.08km), drainage density, Dd (2.40km/km^2), stream frequency, Fs (6.79), drainage texture, Rt (15.39), form factor, Rf (0.279), circulatory ratio, Rc (0.294), elongation ratio, Re (0.596), compactness constant, Cc (1.84), length of overland flow, Lg (0.83) and shape index, Sw (3.58) as listed in **Table 3**. **Figure 5 (c)** shows the topography of the catchment with elevation ranging from 140m to 260m from sea level.

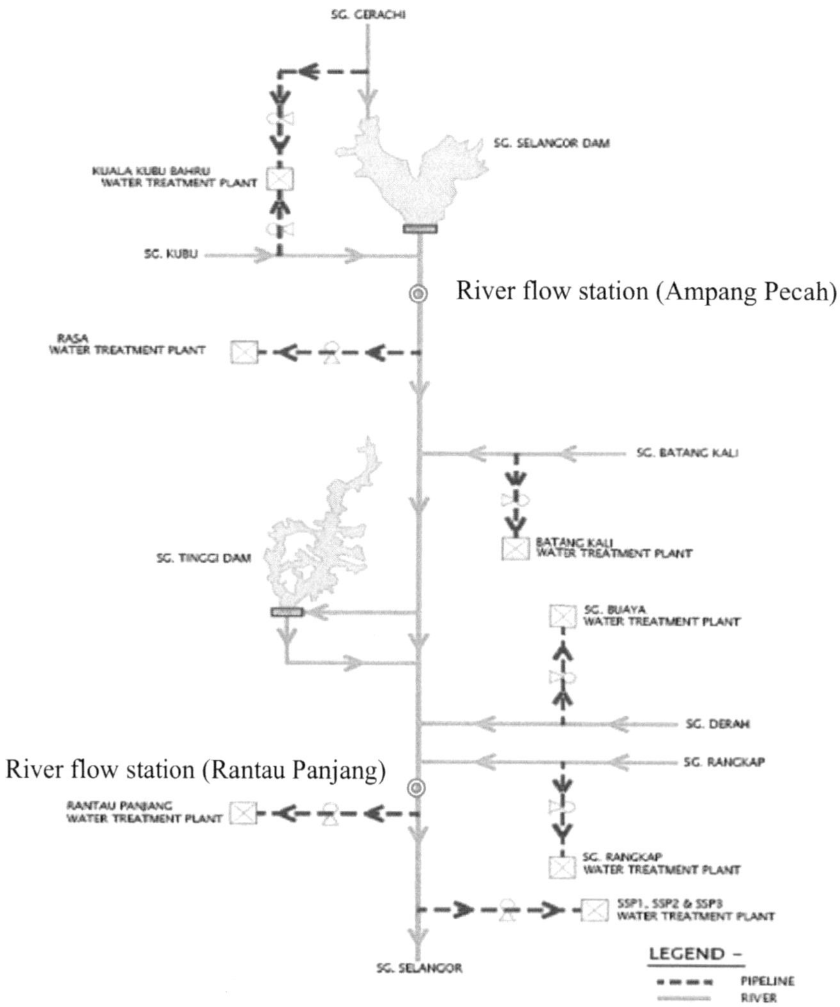

Figure 3. Water intake and dam location.

Table 1. Information of dam.

Dam	Catchment Area (km2)	Storage capacity (m/c/m)	Year completed
Sg Selangor	197	230	2004
Sg Tinggi	40	115	1997

Table 2. Design capacity of water intake.

	Water Intake	Design Capacity (MLD)
1	Sg Tinggi	1.3
2	Kuala Kubu Bharu	6.7
3	Rantau Panjang	31.5
4	Sg Selangor (F1)	950
5	Sg Rasa	250
6	Sg Badong	800
7	Sg Sireh	27
8	Sg Selangor (F2-P1&P2)	950
	Total	3016.5

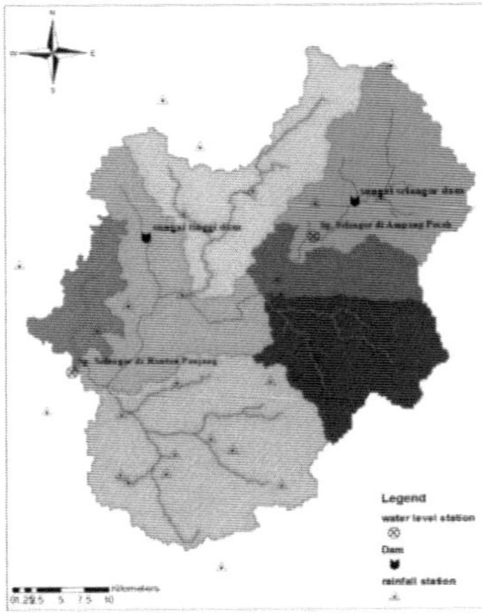

Figure 4 (a): CA

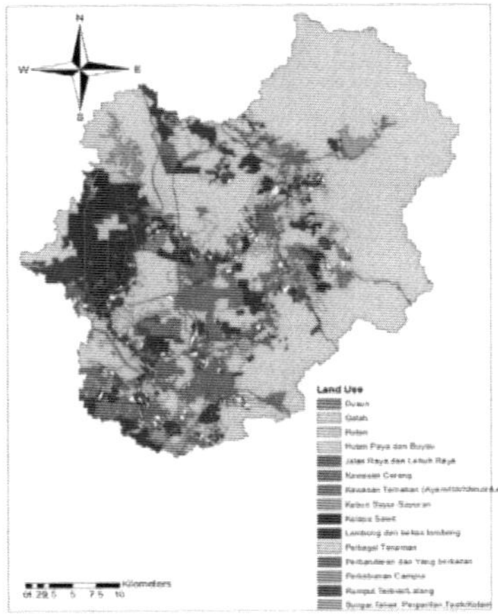

Figure 4 (b): Land Use of CA

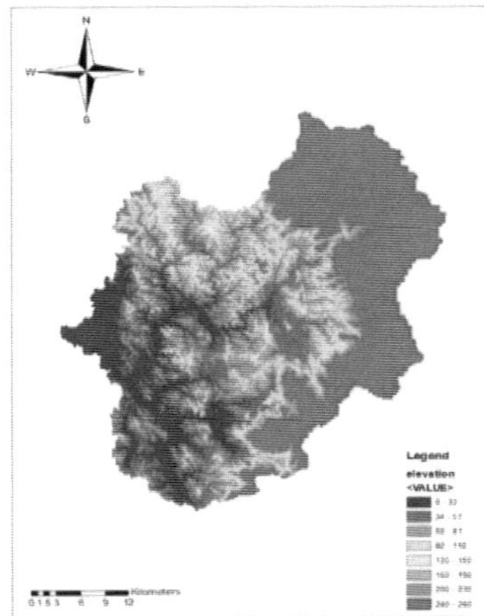

Figure 4 (c): Elevation CA

Figure 4 (d): Drainage CA

Figure 5 (a): CB

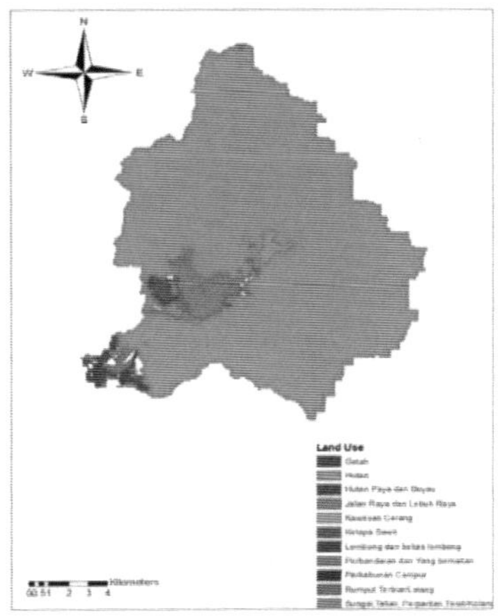

Figure 5 (b): Elevation CB

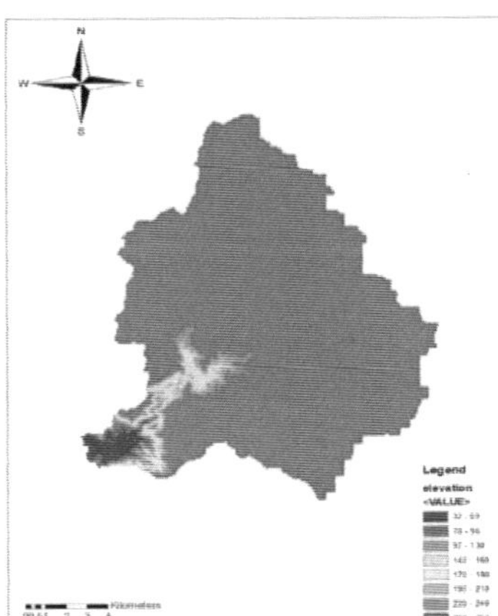

Figure 5 (c): Land Use of CB

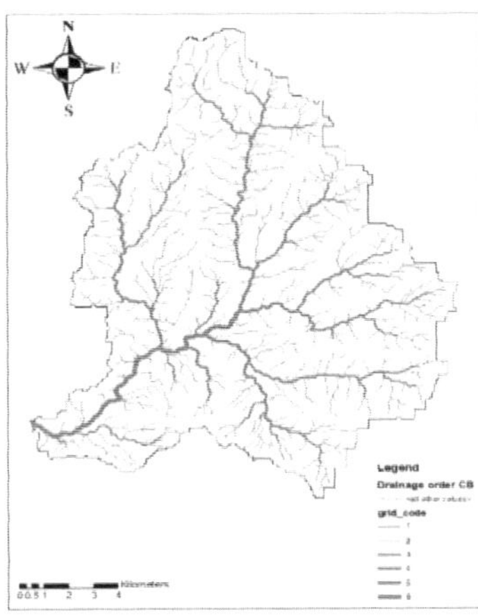

Figure 5 (d): Drainage CB

Figure 5. Catchment information for Catchment A and Catchment B

Table 3. Characteristic of CA and CB.

	CA	**CB**
Area, A (km^2)	1502	220
Land use	Forest (38%), Rubber tree (21.6%), Residential (20.7%), Palm oil tree (9.8%), ex-mining land (6.2%), mangrove tree (3.7%)	Forest (79.3%), Residential (9.8%), Rubber tree (5.4%), Palm oil tree (4.4%), ex-mining land (1.1%),
Elevation	33m – 260m	140m – 260m
Perimeter, P (km)	275	97
Stream order, u	7	6
Total no of stream, Nu	10738	1493
Stream length, Lu	3875	528
Mean stream length, Lsm	553.51	87.96
Basin length, Lb (km)	83.61	28.08
Drainage density, Dd (km/km^2)	2.58	2.40
Stream Frequency, Fs	7.15	6.79
Drainage texture, Rt	39.05	15.39
Form factor, Rf	0.215	0.279
Circulatory ratio, Rc	0.250	0.294
Elongation ratio, Re	0.523	0.596
Compactness constant, Cc	2.00	1.84
Length of overland flow, Lg	0.78	0.83
Shape index, Sw	4.65	3.58

3. Methodology

Rainfall and river flow data were obtained from Department of Irrigation and Drainage (DID) at hourly interval from year 1990 to year 2018. Besides that, the Digital Elevation Model (DEM) data were downloaded from Alos Palsar Alaska Satellite Facility (ASF) with 15m resolution. The DEM data were analyzed using GIS pre-processing (version 10.4.1) as shown in **Figure 6** to determine the watershed characteristics. The details information on the location of water intake, rainfall station, river flow station and dam were obtained from Lembaga Urus Air Selangor (LUAS). Lastly, for land use, data were obtained from Department of Agriculture (DOA) for year 2015.

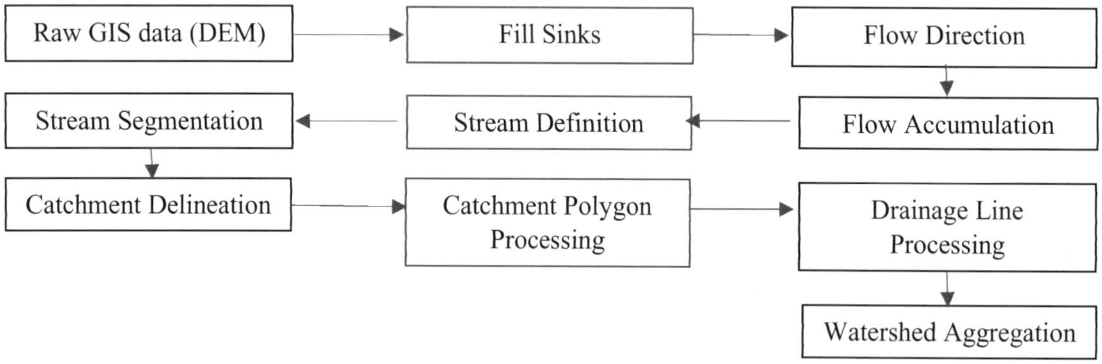

Figure 6. GIS Pre-processing ArcGIS.

The method to study the hydrological responses of land use changes include statistical analysis and model simulation. For statistical analysis, the methods include trend analysis of long time series and the paired catchment approach [11]. In this study, Selangor River Basin (SRB) in Selangor Malaysia, was chosen as the case study to understand the catchment response for different landscapes, geology, catchment characteristics and dam effect on total runoff. To achieve objective 1, which is to calculate the rainfall runoff response for different catchment characteristic, a data set of rainfall and runoff values were used. The data were from year 2007 to year 2017 for Catchment A while data from year 2009 to year 2018 (ten years of analysis) were used for Catchment B. Missing and errors in rainfall data were filled by multiple imputation linear regression method using R program (MICE). Then, Thiessen Polygon method was applied to obtain the areal rainfall for each catchment. There are 27 rainfall stations for Catchment A and three rainfall stations for Catchment B. The method is widely used due to its practicability and less time consuming with highly accurate estimates. After that, the total runoff over total annual ariel rainfall for each year were calculated to obtain the runoff coefficient, c. The runoff coefficient, c, is indicated to measure the rainfall runoff response. The study used trend analysis of long time series to describe the rainfall characteristic. The steps are shown in **Figure 7**.

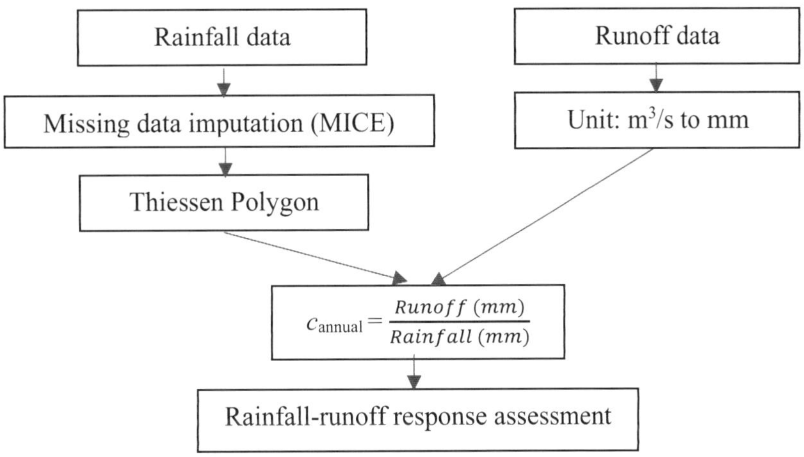

Figure 7. Procedure Rainfall Runoff response.

Beside study the rainfall runoff response within the catchment, the effect of dam on river flow regime
at Selangor River Basin were also investigated. Sungai Selangor dam was completed at year 2004 and Sungai Tinggi dam was completed at year 1997 but the size of Sungai Selangor Dam is three times bigger than Sungai Tinggi Dam. Therefore, the study focuses on Sungai Selangor Dam on river flow regime. Another set of rainfall and runoff data starting from year 1990 to year 2000 were used. Then, similar procedures were applied as shown in **Figure 7**. Finally, the runoff coefficient, c, was calculated for year 1990 to year 2000 and the c value was compared for both catchments. Graphs showing before and after the construction of dam are plotted as in **Figure 11**.

4. Result and Discussion

4.1. Rainfall Characteristic

The rainfall analysis for CA used 11 years data 2007 to 2017, whereas for CB data from 2009 to 2018 (10 years) were used. Missing data were filled in using multiple imputation linear regression method using R program (MICE). The multiple imputation technique involved three steps: i) data imputation, ii) data analysis and iii) data pooling. In the first step, several plausible complete versions of the incomplete data sets are created. Next, in the second step, the different complete version of the incomplete dataset was analyzed using standard statistical procedure which result in multiple different outcomes of the statistical analyses. Finally, these results are combined into the overall statistical analysis where the uncertainty about the missing data is incorporated in the standard errors and significance tests [17].

In CA, 27 rainfall stations were used whereas for CB, only three stations are available. Data from all stations were used to calculate areal rainfall using Thiessen Polygon method in ArcGIS. Thiessen polygon method is widely applied for its accuracy and fast computation. The calculation of Thiessen polygon method is simple because only area of sample data is needed. When dataset is stable, the accuracy of Thiessen polygon method is higher than arithmetic mean [18]. The areal rainfall was calculated for the mean monthly and annual total.

Over 11 years from 2007 to 2018, CA received annual rainfall in a range from 2207mm to 3071mm with mean annual of 2699mm/yr. In CB, the annual rainfall ranges from 1888mm to 3461mm with mean annual of 2507mm/yr. Like CA, the driest and wettest years in CB were in 2016 and 2012, respectively. The driest in 2016 could be due to the El Nino event, where it was found that Malaysia was strongly effected by El Nino in the years from 2014 to 2016 resulting to abnormal amount of rainfall [19]. The trend of rainfall for both catchments shows decreasing trend in the total annual rainfall received. **Figure 8** shows a comparison of annual rainfall between the two catchments.

Both catchments show that November received higher rainfall (333 mm/month) while February was the driest with 122 mm/month. The variation of rainfall per month at CA and CB is generally characterized by bimodal variations with two peaks as shown in **Figure 9** which are primary peak occurring during November and secondary peak during April or May due to inter-monsoon periods. During the monsoon transitions (inter-monsoon) periods, the west coast of Peninsular Malaysia receives more rainfall compared to other seasons [14]. This is because the weather in CA and CB are influenced by two monsoon regimes, which are northeast monsoon (NEM) (known as wet season) that brings heavy rainfall in Malaysia particularly to the east coast states of Peninsular and southwest monsoon (SWM) (known as dry season) normally with relatively drier weather. Record shows that widespread floods have occurred especially during the NEM season [12].

The rainfall usually pours at the evening from 3pm to 10pm (**Figure 10**) for all months for both catchments. The rainfall is a convective rain that occurs when the energy of the sun at the noon heats the surface of the earth causing water to evaporate to form water vapor. Convective rainfall is widespread in areas where the ground is heated by the hot sun such as the tropics [20]. Therefore, Malaysia's rainforests experience heavy rainfall mostly at the afternoons.

4.2. Rainfall Runoff Response

The rainfall-runoff analysis at Rantau Panjang was conducted to assess the dam influence on the streamflow regime. **Figure 11** shows the rainfall-runoff relationship for CA and CB. Prior to dam construction, the annual runoff generally increases as rainfall increases for both catchments. However, after the dam construction, the runoff in CB decreases when rainfall increases. This pattern is not expected under normal condition. The dam was constructed to store as much water as possible during rainy season and to be used during dry periods, hence the rainfall-runoff relationship for CB is highly dependent on the dam operations and is not influenced by natural catchment response.

The effect of dam however was not significant for CA other than a slight change in the slope of the rainfall-runoff relationship and the runoff means. Before the dam construction, the total average runoff for CA was lower (896 mm) compared to after construction (1129 mm). This is because most of the area at Selangor River Basin (SRB) were still pervious in year 1990 to year 2000. Therefore, the infiltration rate is higher. Expended impervious surfaces such as parking lots, roofs, driveways and sidewalks will block the precipitation infiltrate into the groundwater thus increase the total volume and peak discharge of the streamflow [21]. Rainfall falling on impervious areas is presumed to produce surface runoff. Increasing the percentage of impervious area would result to increase of the runoff volume, reduced flow time and lag time, increased in peak discharges with resulting shift in the flood frequency curve [22].

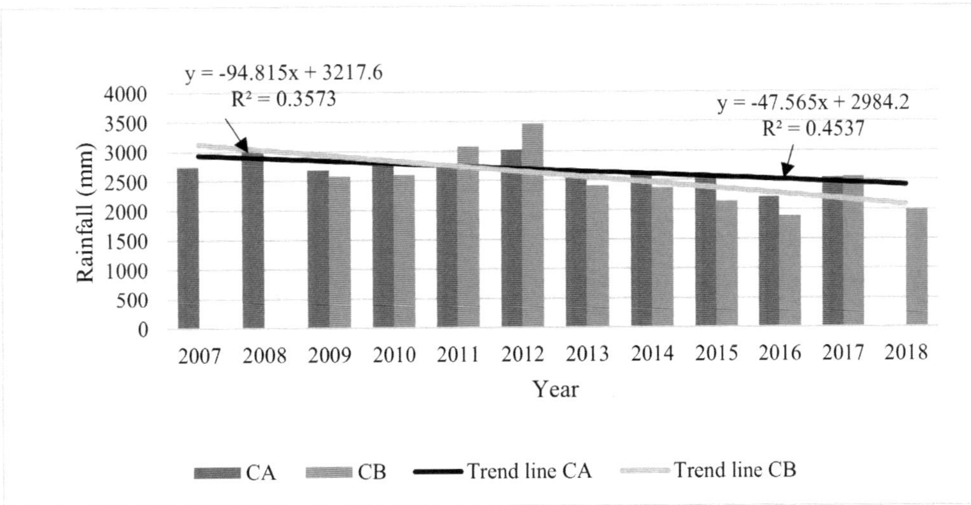

Figure 8. Annual rainfall CA and CB.

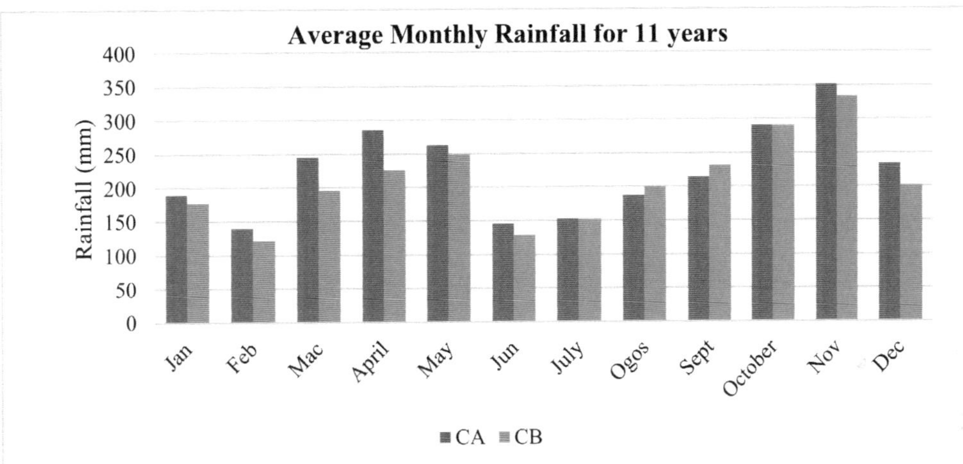

Figure 9. Monthly rainfall CB and CB.

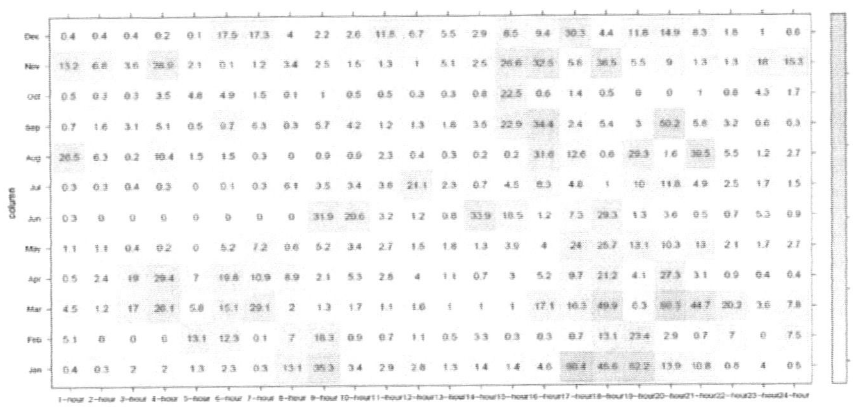

Figure 10. Heat map for CA and CB.

Before dam construction, the rainfall runoff response for CB was higher compared to CA (**Figure 11**). The volume of runoff from CB is higher than CA for any rainfall events. This may be due to larger water catchment storage in CA and the amount of water intakes' facilities within the catchment. Catchment A has eight water intake stations in total within the catchment compared to CB which has only one. Therefore, the runoff water in CA was extracted thus reducing the amount of runoff volume extracted from the observed hydrographs. Besides that, morphometric perimeter for both catchments is also different. Circulatory ratio, Rc, and elongation ratio, Re for CB is higher than CA meaning CA indicates low relief and impermeable surface thus lower lag time compared to CB. Catchment B also have strong relief and steep slopes, meaning shorter time to peak. The form factor, Rf, for CB is larger than CA, thus CB has high peak flow of short duration compared to CA.

After the construction of dam, the rainfall-runoff model was not significant for Catchment B which has a p value of 0.439 (p-value > 0.05) as shown in **Table 4.** This is due to the regulated water from Sungai Selangor dam consequence to insignificant relationship between the observed rainfall and runoff. According to Adam et al. (2007), the construction and operation of large reservoir results in shift in streamflow seasonality that reduces peak flows and increases low flows. Another effect of dams regulation is reducing peak flow during the flooding season and releasing during the dry season due to change in the river flow regime [10]. **Figure 13** shows the flow duration curve for CA and CB before and after dam construction. Slight changes of the high and low flows are seen for CA, however for CB some significant changes for the high flows can be seen. The high flows for CB were reduced after the dam construction. This correlates with the dam operation of reducing the runoff volumes particularly during high flows.

Table 4. Model Performance and Evaluation.

	CA before dam construction	**CB before dam construction**	**CA after dam construction**	**CB after dam construction**
R^2	0.88	0.83	0.75	0.08
p-value	0.0018	0.004	0.00058	0.439

Assessment on the runoff coefficient, *c*, is shown in **Figure 12**. For CA, before dam construction the c values are between 0.19 to 0.48 indicating that only 19% to 48% of rainfall turned into runoff but after the existence of dam, the *c* value changed within the range of 0.35 to 0.49, **Table 5**. This shows an increase of the lower range of the runoff coefficients.

For CB, before construction of dam, the *c* values were between the range of 0.59 to 0.79, but then changes to the range of 0.28 to 0.89 after dam construction. However, as from CA, the lower range of *c* values reduce to 0.28 from 0.59 and larger values of *c* increase from 0.79 to 0.89. This indicates the variability of the rainfall-runoff response becomes larger for CB. This is due to the regulated water from the dam, where during dry season, the dam releases water into the river and vice versa during wet season. Then, runoff coefficient, *c*, for CB is as shown in **Figure 12**.

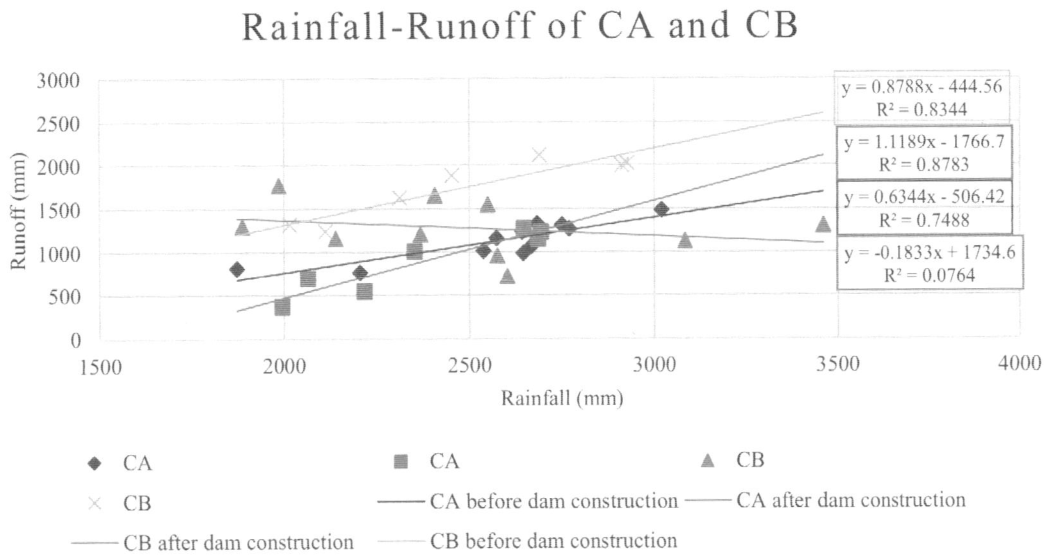

Figure 11. Rainfall-Runoff Relationship.

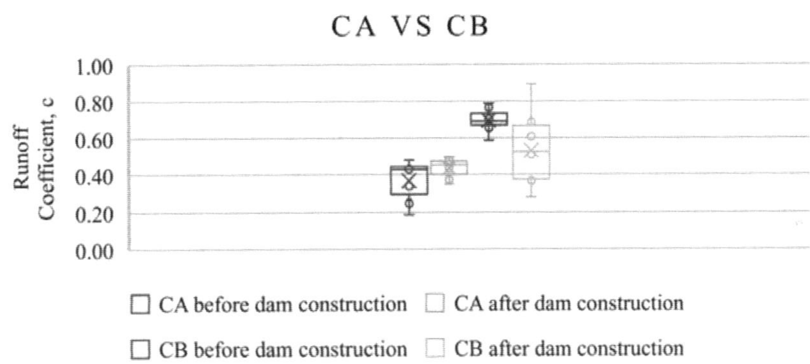

Figure 12. Boxplot of Runoff Coefficient, *c* for CA and CB.

Table 5. Runoff Coefficient, *c*.

	CA prior to dam construction	CA after dam construction	CB prior to dam construction	CB after dam construction
min	0.19	0.35	0.59	0.28
Q1	0.29	0.40	0.67	0.37
median	0.43	0.45	0.69	0.53
Q3	0.44	0.47	0.73	0.67
max	0.48	0.49	0.79	0.89
mean	0.37	0.44	0.70	0.53
range	0.29	0.15	0.20	0.61

IOP Conf. Series: Materials Science and Engineering 1153 (2021) 012004 doi:10.1088/1757-899X/1153/1/012004

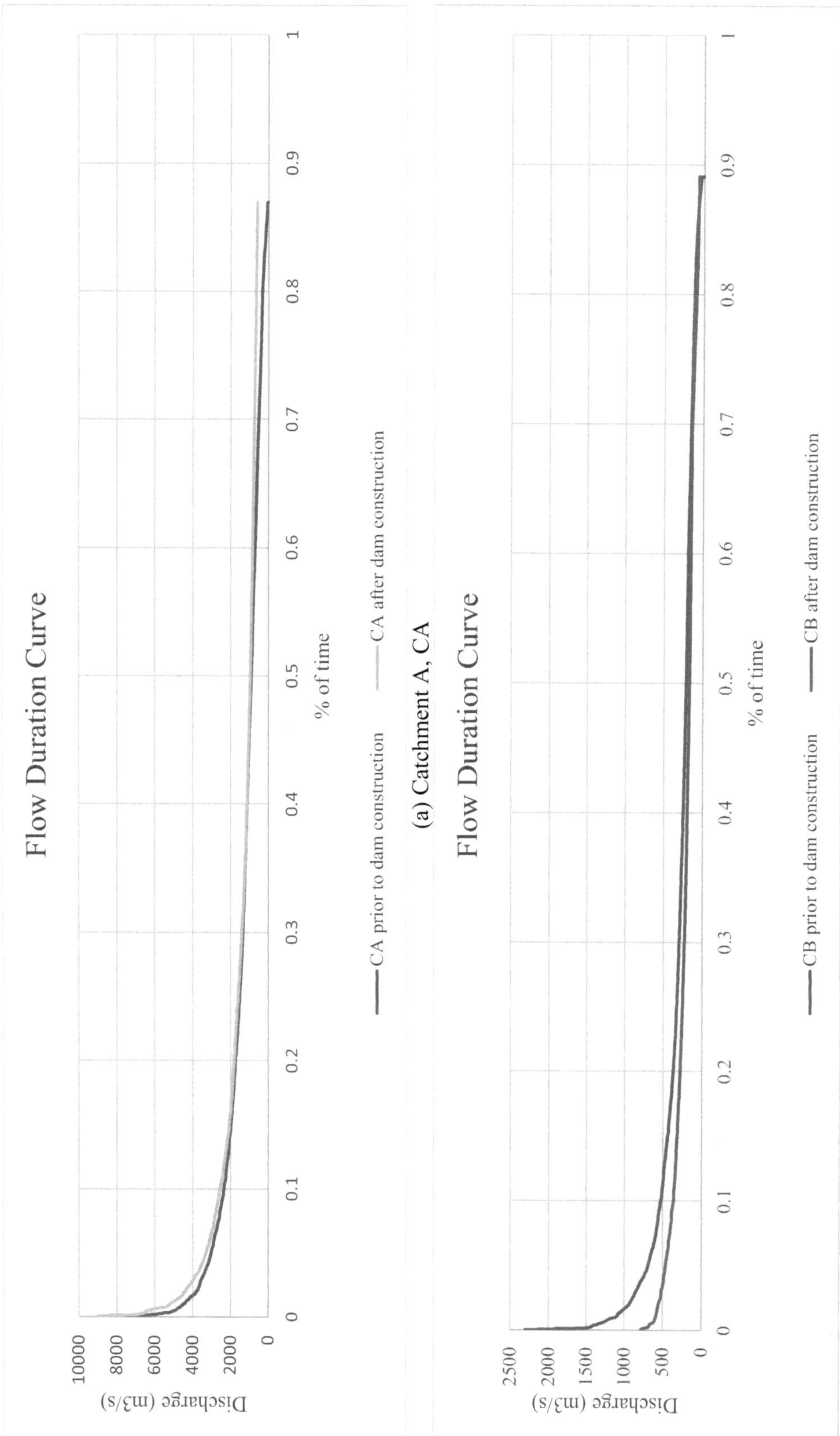

Figure 13: Flow Duration Curve

5. Conclusion

The paper assessed the rainfall-runoff response of two catchments with different characteristics. One is the Sungai Selangor River Basin as Catchment A (CA) and another, a sub-catchment of CA located at the upstream, as Catchment B (CB). Both catchments have similar exposure of rainfall. The total rainfall has a decreasing trend for the past 10 years, and the monthly rainfall was characterized by bimodal variation: first peak in November and second peak in April or May due to effect of inter-monsoon. Both catchments were also affected by Sungai Selangor Dam operation with CB having the most significant changes. Catchment B runoff coefficient, *c* and runoff were higher before and after the dam construction compared to CA. The variation of *c* for CB becomes higher after the dam construction, while the *c* values for CA only changes slightly for the lower range of *c* values. This is because the dam regulates water during wet and dry season. As the dam is in CB, CB flow and runoff changed significantly. By understanding the runoff coefficient, we can learn how the catchment characteristics can transform rainfall into runoff and provide a basis for information about the catchment response. The result also shows that, Sungai Selangor Dam regulates the water flow regime by reducing the peak flow during flooding season and releasing during the dry season, hence changing the high and low flows at the catchment outlet.

Acknowledgement

The authors would like to state their appreciation to the funding from the Malaysia Research University Network (MRUN), vot number R.J130000.7851.4L897. The authors also thank the Department of Irrigation and Drainage Malaysia (DID), and the Selangor Water Management Board (LUAS) for providing data and information for the research.

References

[1] F. Othman, M. S. U. Chowdhury, W. Z. Wan Jaafar, E. M. M. Faresh, and S. M. Shirazi, "Assessing risk and sources of heavy metals in a tropical river basin: A case study of the Selangor river, Malaysia," *Polish J. Environ. Stud.*, vol. 27, no. 4, pp. 1659–1672, 2018.

[2] A. Balasubramanian and D. Nagaraju, "The Hydrologic Cycle," *Int. Geophys.*, vol. 56, no. C, pp. 115–135, 1994.

[3] M. S. Johnson, W. F. Coon, V. K. Mehta, T. S. Steenhuis, E. S. Brooks, and J. Boll, "Application of two hydrologic models with different runoff mechanisms to a hillslope dominated watershed in the northeastern US : a comparison of HSPF and SMR," vol. 284, pp. 57–76, 2003.

[4] W. Yang, D. Li, T. Sun, and G. Ni, "Saturation-excess and in fi ltration-excess runoff on green roofs," *Ecol. Eng.*, vol. 74, pp. 327–336, 2015.

[5] T. Blume, E. Zehe, and A. Bronstert, "Rainfall-runoff response, event-based runoff coefficients and hydrograph separation," *Hydrol. Sci. J.*, vol. 52, no. 5, pp. 843–862, 2007.

[6] D. B. Thompson, "The Rational Method," *Eng. Hydrol.*, no. January, p. 21, 2006.

[7] J. Liu, C. Zhang, L. Kou, and Q. Zhou, "Effects of Climate and Land Use Changes on Water Resources in the Taoer River," *Adv. Meteorol.*, vol. 2017, 2017.

[8] M. Burke, K. Jorde, and J. M. Buffington, "Application of a hierarchical framework for assessing environmental impacts of dam operation: Changes in streamflow, bed mobility and recruitment of riparian trees in a western North American river," *J. Environ. Manage.*, vol. 90, no. SUPPL. 3, pp. S224–S236, 2009.

[9] H. Wang, F. Sun, and W. Liu, “Characteristics of streamflow in the main stream of Changjiang River and the impact of the Three Gorges Dam,” *Catena*, vol. 189, no. February, p. 104498, 2020.
[10] Y. Yan, Z. Yang, Q. Liu, and T. Sun, “Assessing effects of dam operation on flow regimes in the lower Yellow River,” *Procedia Environ. Sci.*, vol. 2, no. 5, pp. 507–516, 2010.
[11] P. Li, H. Li, G. Yang, Q. Zhang, and Y. Diao, “Assessing the hydrologic impacts of land use change in the Taihu Lake Basin of China from 1985 to 2010,” *Water (Switzerland)*, vol. 10, no. 11, 2018.
[12] J. Suhaila, S. M. Deni, W. A. N. Zawiah Zin, and A. A. Jemain, “Trends in Peninsular Malaysia rainfall data during the southwest monsoon and northeast monsoon seasons: 1975-2004,” *Sains Malaysiana*, vol. 39, no. 4, pp. 533–542, 2010.
[13] C. L. Wong, J. Liew, Z. Yusop, T. Ismail, R. Venneker, and S. Uhlenbrook, “Rainfall characteristics and regionalization in peninsular malaysia based on a high resolution gridded data set,” *Water (Switzerland)*, vol. 8, no. 11, 2016.
[14] M. F. Mohd Akhir, N. Z. Zakaria, and F. Tangang, “Intermonsoon Variation of Physical Characteristics and Current Circulation along the East Coast of Peninsular Malaysia,” *Int. J. Oceanogr.*, vol. 2014, pp. 1–9, 2014.
[15] S. Nurhidayu and M. Azhar, “Long-Term Sediment Pattern of the Selangor River Basin , Malaysia Impacted By Land-Use and Climate Changes,” no. October, pp. 1995–1998, 2015.
[16] LUAS, “Sungai selangor basin management plan 2015 – 2020,” pp. 1–346, 2014.
[17] J. R. van Ginkel, M. Linting, R. C. A. Rippe, and A. van der Voort, “Rebutting Existing Misconceptions About Multiple Imputation as a Method for Handling Missing Data,” *J. Pers. Assess.*, vol. 102, no. 3, pp. 297–308, 2020.
[18] D. Han and M. Bray, “Automated Thiessen polygon generation,” *Water Resour. Res.*, vol. 42, no. 11, pp. 2–6, 2006.
[19] Nur Nadia Kamil and S. F. Omar, “The Impact of El Niño and La Niña on Malaysian Palm Oil Industry,” *Oil Palm Bull.*, vol. 74, no. May 2017, pp. 1–6, 2017.
[20] C. M. Alcala and A. E. Dessler, “Observations of deep convection in the tropics using the Tropical Rainfall Measuring Mission (TRMM) precipitation radar,” *J. Geophys. Res. Atmos.*, vol. 107, no. 24, 2002.
[21] C. Zhu and Y. Li, “Long-Term Hydrological Impacts of Land Use/Land Cover Change From 1984 to 2010 in the Little River Watershed, Tennessee,” *Int. Soil Water Conserv. Res.*, vol. 2, no. 2, pp. 11–21, 2014.
[22] J. Du *et al.*, “Assessing the effects of urbanization on annual runoff and flood events using an integrated hydrological modeling system for Qinhuai River basin, China,” *J. Hydrol.*, vol. 464–465, pp. 127–139, 2012.
[23] J. C. Adam, I. Haddeland, F. Su, and D. P. Lettenmaier, “Simulation of reservoir influences on annual and seasonal streamflow changes for the Lena, Yenisei, and Ob’ rivers,” *J. Geophys. Res. Atmos.*, vol. 112, no. 24, pp. 1–22, 2007.

IOP Conf. Series: Materials Science and Engineering 1153 (2021) 012005 doi:10.1088/1757-899X/1153/1/012005

Feasibility of low impact development measures to mitigate inundation in tidal – impacted urban area: A case in Ho Chi Minh City, Vietnam

N Pham[1], A Cabaltica[1], H T Pham[2]

[1]Department of Civil Engineering, International University – Vietnam National University Ho Chi Minh City, Vietnam
[2]School of Biotechnology, International University – Vietnam National University Ho Chi Minh City, Vietnam
Email: pngoc@hcmiu.edu.vn

Abstract. Recently, urban inundation is becoming more severe in Ho Chi Minh City. Many structural measures have been proposed, but low impact development (LID) measures have not been sufficiently considered. This study quantitatively investigated the possibility of LID measures in District 12, Ho Chi Minh City using modelling approach. Based on the collected secondary data on land use, sewer network, and terrain elevation, a semi-two-dimensional storm water management model (PCSWMM-2D) was constructed and calibrated against heavy rainfall and high tidal event on 26 September 2016. Errors between measured and simulated inundated depths at several positions was smaller than 5%; thus, the calibrated model was at reliable accuracy to predict design scenarios. Using 10-year return period scenarios of tidal levels at Sai Gon River and rainfall at Tan Son Hoa gauging station, the model was used to analyse several designed LID measures including green roof, pervious pavement, and storage or flood retention lake/pond. The simulated results indicated that if green roof and pervious pavement are separately applied, the average magnitude of inundation depth can be reduced by 11.4% and 14.0%, respectively. Meanwhile, when a well-designed storage system including a gated-flood retention lake in combination with suitable conduits and a pumping station is constructed, the flooding situation is completely controlled. In conclusion, it is suggested that to manage urban inundation in terms of sustainable development for Ho Chi Minh City, a well-designed storage system is needed prior to the urbanizing area. In addition, those LID measures should be widely applied as long-term strategy for flooding management in Ho Chi Minh City.

1. Introduction

Ho Chi Minh City (HCMC) is the biggest city of Vietnam located in lower Dong Nai river basin with the total area of 2.061 km^2. The city is characterized as a lowland area with dense river network. There are two big rivers (namely, Dong Nai river and Sai Gon river) together with many tributaries flowing throw the city (Figure 1); which is estimated to have 975 km length used for navigation. This area is a relatively flat and low-lying terrain. The regions with the elevation (AMSL) less than + 0.5 m and less than + 1.0 m are 18.67% and 53.54%, respectively of the total area and are mostly located along the rivers or canals in the city. Additionally, the city also is influenced by the semi-diurnal tide of the East Sea. Recently, under the context of climate chance and sea level raising, frequency of high tidal water levels is recently increasing; therefore, the drainage situation of the city becomes more difficult leading to more tidal flooding in the city [1]. For example, as the result of high tidal phenomena occurred on 30 September 2019 with the peak of water level rising up to + 1.77 m at Phu An gauging station in Sai Gon river, a large area in the south of the city is highly inundated and prolonged.

Moreover, HCMC is in the tropical monsoon region with a high average annual rainfall ranging from 1,200 mm to 1,900 mm. Nevertheless, ninety percent of this annual value mainly occurs during the rainy season (from May to November). On the other hand, intensity and frequency of heavy rainfall events caused city flooding tend to increase in recent years [2]. Consequently, urban flooding usually happens more frequently and widespread during the rainy season. The flooding situation becomes more severe when heavy rains and high tides simultaneously occur. For instance, a heavy storm on November 25, 2018 from 07:00 to 15:00 produced 401 mm rainfall measured at Tan Son Hoa gauging station. The peak tidal level measured at Phu An gauging station at 18:30 on the same day was + 1.29 m, 9 cm higher than flooding alarm level 1 of the city. This resulted to 102 roads in the city being inundated with depths varying from 10 cm to 70 cm [3].

Presently, the flooding situation in the city is still very serious despite many projects implemented under the urban drainage master plan Decision 752 QĐ/TTg [4] signed by the Prime Minister on June 19, 2001; and many previous studies related to urban flood management in HCMC prioritizing conventional hydraulic structures. However, there are few studies considering Low Impact Development (LID) approach. Some of the studies on applying LID measures in HCMC modeled using Storm Water Management Model developed by United States Environmental Protection Agency (EPA-SWMM) are those of Ho et al. [5] and Quan et al. [6]. Ho et al., assessed four Sustainable Urban Drainage Systems (SUDS) namely rainwater harvesting, green roofs, urban green space, and pervious pavement, with application to the Nhieu Loc - Thi Nghe Basin, located in the central part of HCMC. Quan et al. used Tuflow and SWMM models for simulating surface and drainage water system in assessing the effectiveness of green infrastructure development in urban flood reduction. The models were developed at a specific study area namely Tham Luong - Ben Cat Catchment in HCMC. For a more comprehensive discussion on the use SWMM to model the long-term performance of specific types of LID alternatives, which could be referred to the work of Rossman [7].

In line with sustainable flood management for HCMC, the objective of this study is to investigate the feasibility of implementing LID measures, modeled using the Personal Computer Storm Water Management Model (PCSWMM) through a case study in the District 12. The study area is in the north of the city (Figure 1), which is selected due to its high urbanization rate. No studies on application of LID measures for inundation reduction have been conducted in this area yet.

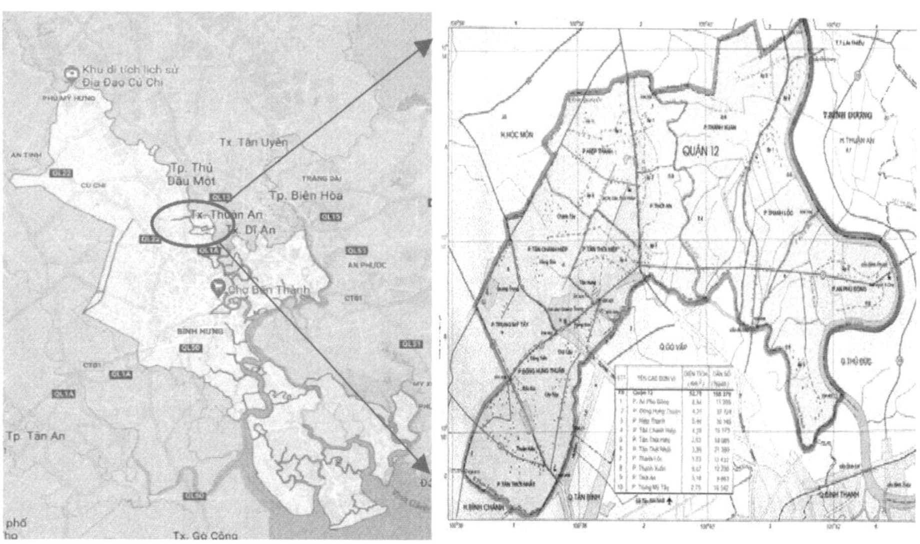

Figure 1. Study area: District 12, Ho Chi Minh City.

2. Methodology and data collection

In this study, the numerical model (PCSWMM model), is used as a supporting tool. PCSWMM is a commercial software conducted by Canadian Computational Hydraulics International. Its fundamental concepts are the same with EPA-SWMM model [8]. In application of the model, two important steps have been carried out, including: (i) calibrating PCSWMM model; and (ii) evaluating the LID measures using the calibrated model. The general research framework is presenting in the figure 2.

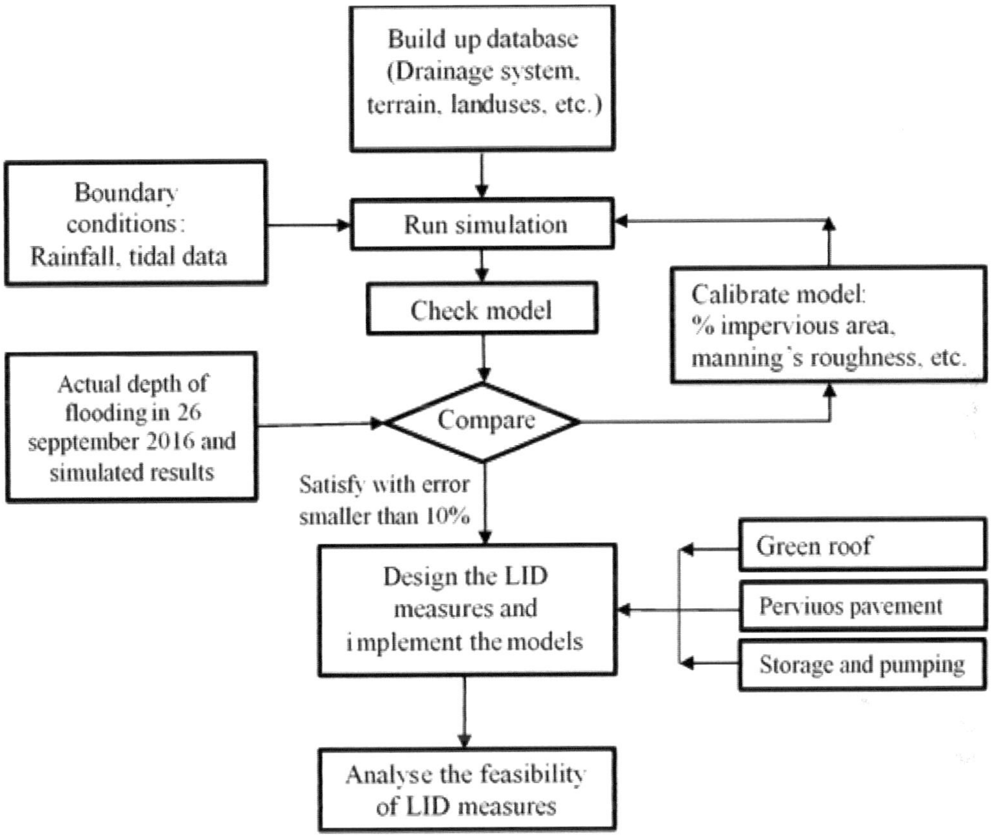

Figure 2. The research framework using PCSWMM model for the case study.

To calibrate PCSWMM model, a huge of data set was officially collected from governmental agencies and academic institutes of Ho Chi Minh City. For example, the map of existing urban drainage system of the city was collected from Department of Construction. The land use and soil map were provided by Department of Natural Resources and Environment. The data on digital elevation map (DEM), climatic and hydrology was transferred from Ho Chi Minh City Institute for Meteorology, Hydrology and Climate Change. The historic flooding water depths in the city were collected from Operations Center of the City Flood Control Program of Ho Chi Minh City, and so forth.

The drainage system of the case study was modelled by some components such as sub-catchment, node (or junction), and culvert (or conduit). Each component had many parameters that were determined directly or indirectly. After standardizing the data, inputting data and calibrating the model have been processed. Following paragraphs is briefly describing how to input some important parameters.

For sub-catchments, the inputted data include percentage of impervious area, slope, maximum and minimum infiltration rates. Percentage of impervious area were taken as 75% to 85% for densely populated areas; and 35% for sub-basins with low levels of urbanization, mostly in the eastern region. The slope of the sub-catchment was based on DEM elevation of District 12 varying from 0.1 to 0.5. The infiltration equation used was Horton's formula with parameters determined for loamy soil (according to soil data), including maximum water infiltration rate and minimum infiltration rate on saturated soil taken as 76.2 mm/h and 7 mm/h, respectively. In addition, decay constant is inputted as 4 L/h; and soil time from saturation to completely dry was 5 days.

For the collection nodes, beside of geometric data, the inflow parameter is defined as amount of wastewater that the nodes are received. This amount of wastewater was calculated basing on domestic water consumption within the sub-basin, according to Decree No. 80/2014/ND-CP signed by the Prime Minister on 6 August 2014 [9], and other national regulations related to water use.

For the conduits, the Manning roughness coefficient, n was selected as 0.013. Moreover, the slope of the conduit was taken as 1/D in which D is diameter of the conduit in cm, following QCVN 07-2:2016/BXD [10]. Other parameters included in the model were based on existing data on drainage system and topographic map.

Moreover, referenced to the previous studies ([5] and [6]) and our experience in flood management for the city, three potential LID measures were considered in the case study, including: (i) green roof, (ii) pervious pavement, and (iii) storage system (flood retention lake/pond and pumping station). Furthermore, evaluating the feasibility of the measures was based on reduction of flooding depth simulated by the calibrated model.

3. Results and Discussion

3.1. Performance of calibrated model

As the result of calibration, the final PCSWMM model for the study area consisted of 168 sub-catchments, 267 nodes (junctions) and 319 culverts (conduits) as showing in the figure 3. Moreover, the data of hydrological boundary conditions were rainfall intensity and water levels on rivers and canals at the outlets of existing sewer systems (or outfalls) were inputted in the model. A real storm event, with estimated return period of 10 years which occurred on 26 September 2016, was selected for calibration. The design rain pattern was determined using the IDF curve provided by the Ho Chi Minh City Institute for Meteorology, Hydrology and Climate Change. The water levels at the outfalls located in Vam Thuat river and Sai Gon river were extracted from a Mike 11 model constructed for lower Dong Nai river system from a different project by the author [11]. Those inputted parameters are shown in the figure 4.

Figure 3. Model setup in PCSWMM software.

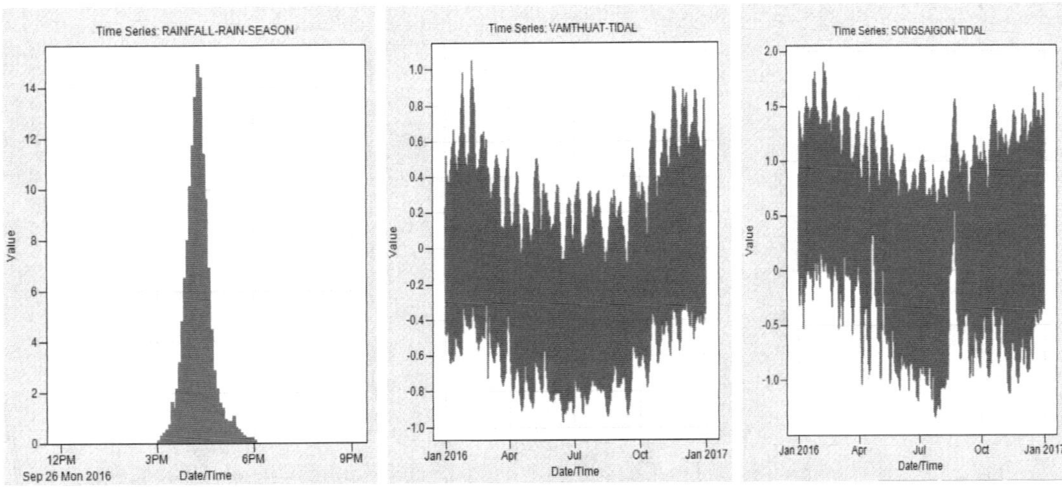

Figure 4. Rainfall data and two tidal curves of Vam Thuat river and Sai Gon river.

To calibrate the model, data of inundated depth measured at four locations after the rain on 26 September 2016 was used. The results showed that the error between simulated and measured depths are less than 10% as illustrated in the figure 5, which is indicated that the calibrated model is good reliability for predicting designed scenarios. The calibrated model then was applied for three different scenarios employing LID measures namely green roof, permeable pavement, and retention lake with pumping station to assess their potential in inundation reduction.

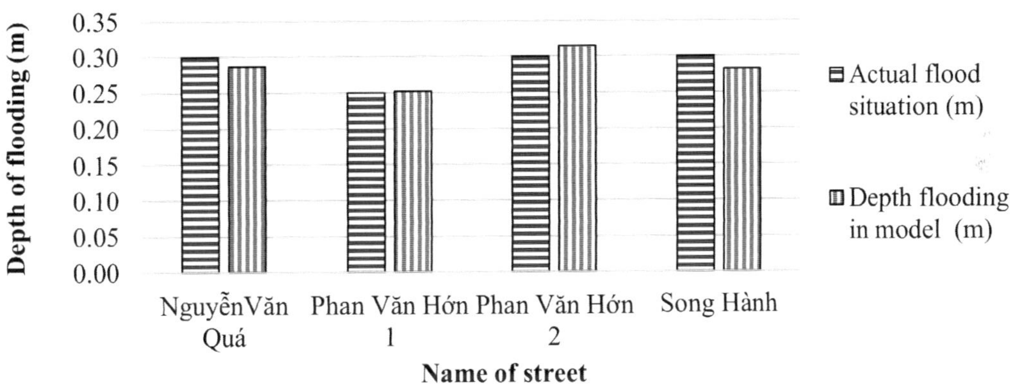

Figure 5. Calculated and measured inundation depths at different locations in District 12.

3.2. Evaluation of LID measures in inundation reduction

3.2.1. Green roof

Some areas in the western part of the district that have been highly urbanized was assigned an impermeability of 63% (including residential areas, industrial parks, services, etc.) in the basic model (without LID measures applied). Potentially, to reduce inundation, these areas should be converted into green roof, especially for the region along Nguyen Van Qua Street (green circle in the figure 6). Based on technical design of the green roof planted by dense vetiver grass (Figure 6a), parameters of the roof were inputted for the model in 3 categories such as surface, soil, and drainage mat. The simulated results show that, if the green roof measure is adapted for all the buildings having good structure (e.g.: high residential and commercial buildings, industrial buildings and so on) in this sub-catchment, the flooding level can be reduced by as much as 17% for the scenario with rain and high tide with probability of occurring P = 10% (Figure 7).

Figure 6. The regions where (a) green roof and (b) permeable pavement were investigated in the PCSWMM model.

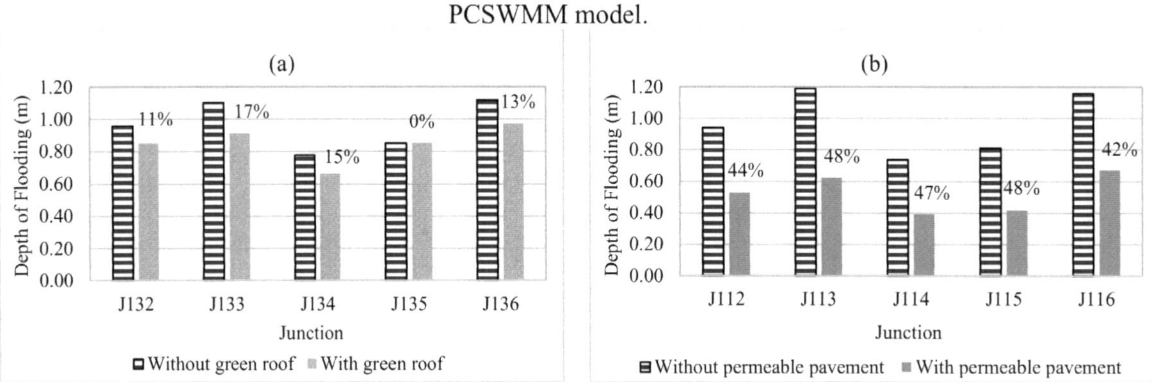

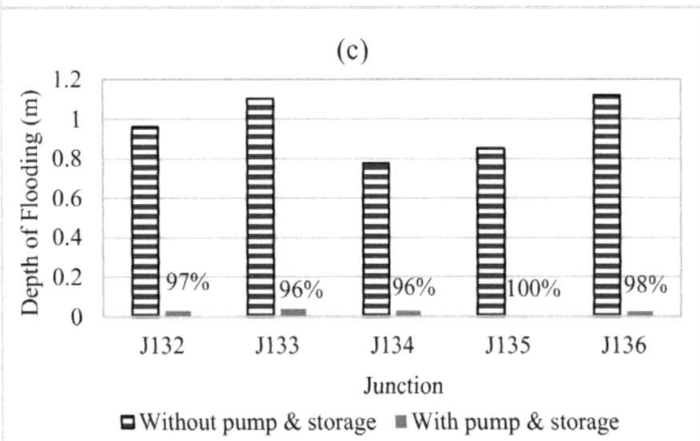

Figure 7. Flood depth simulation results with and without using LID measures: (a) green roof, (b) permeable pavement, and (c) retention lakes and pumping station. Values represent percent reduction in depth of flooding.

3.2.2. Permeable pavement

The geological condition of District 12 is old alluvial soil mixed with sandy loam soil in the western part which has high level of urbanization, and clay mixed with impurities in the eastern part with the dense canal system. Therefore, increasing surface permeability appropriate for each sub-region has potential in reducing floods. Permeable pavement is applied for Song Hanh street (blue circle in figure 6). The permeable pavement was designed for this area showing in the figure 6b, which are several permeable lavers in vertical direction, including block paver, porous concrete (or asphalt mix), and gravel with drainage pipe system. Based on the designed pavement, numerical data is standardized for the model. The modelling results show that permeable pavement can reduce the amount of flooding in the surrounding area by as much as over 40% (Figure 7b).

3.2.3. Retention lake and pumping station

The areas along Phan Van Hon and Nguyen Anh Thu Street have enormous spaces to build a flood retention lake, the red circle in the figure 8a. To control the flooding along those streets, a system was designed, which consists of a retention lake with a volume of 40,000 m^3 together with gated-pump station as illustrated in the figure 8b. The pump with the head above the head of junction was chosen. The pump will be operated when drainage capacity of sewer system is over as the result of very heavy rainfall or tidal water level higher than probability design. Based on modelling results, this measure shows significant reduction in inundated depth (Figure 7c), which is various from 96% to 100%. Comparing to the other proposed measures, this one is the best measure to reduce urban flooding in the district 12.

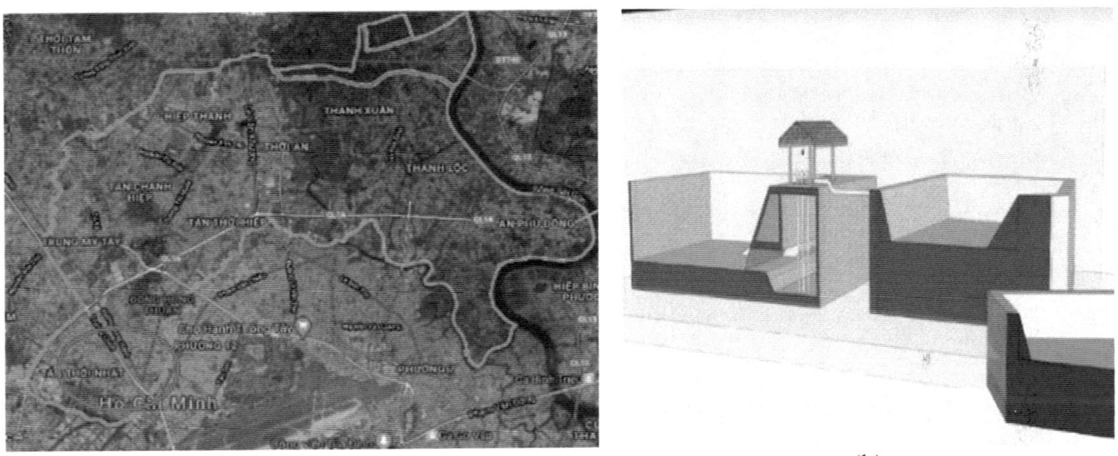

(a) (b)

Figure 8. (a) Location of the lake and pumping station in the map of District 12, (b) system design.

4. CONCLUSION

With aid of PCSWMM model, feasibility of potential LID measures; which were green roof, permeable pavement and a retention lake together with pumping station, are investigated through a case study in the District 12, Ho Chi Minh City. The simulated results of inundation depth reduction show that a well-designed storage system including a gated-flood retention lake in combination with suitable conduits and a pumping station is the best measure. Therefore, this measure is suggested as priority to build at urbanizing area of the city because its land occupation. In addition, designed green roof, and pervious pavement are not much effective in flooding reduction, estimated as 11.4% and 14.0% of inundated depth, respectively. However, application of these two measures also is encouraged as long-term strategy for flooding management in the city because of some benefits such as improving the city landscape by green space, discharging underground water, and so on.

Acknowledgments
This research was supported by Department of Scientific and Technology of Ho Chi Minh City under the project entitled "Conducting the map of pollutant propagation of surface water resources in the

city", implementing from 2017 to 2020. We also thank our colleagues from Ho Chi Minh City International University who provided insight and expertise that greatly assisted the research.

References

[1] Quang C N X 2015 Analysing the tendency of annual peak water level in lower Dong Nai – Sai Gon river basin *Journal of Agricultural and Rural Development* **Vol.2** pp 65–71.
[2] Ngoc P 2019 Propose sustainable rainwater management for Ho Chi Minh City *Builders Journal* **7&8** pp 30–35.
[3] Annual Report 2018 Center of inundation protection of Ho Chi Minh City www.ttcn.hochiminhcity.gov.vn.
[4] Approval of the general planning of the drainage system of Ho Chi Minh City by 2020 *Prime Minister's Decision No: 752 / QD-TTg.*
[5] Ho H L, Babel M S, Weesakul S, Irvine K N, and Pham M D 2015 Exploratory assessment of SUDS feasibility in Nhieu Loc-Thi Nghe Basin, Ho Chi Minh City, Vietnam *British Journal of Environment and Climate Change, 5(2)* pp 91-103.
[6] Quan N H, et al. 2019 Green Infrastructure Modelling for Assessment of Urban Flood Reduction in Ho Chi Minh City *CIGOS Innovation for Sustainable Infrastructure Springer, Singapore* **vol 54**, eds Ha-Minh C, Dao D, Benboudjema F, Derrible S, Huynh D, Tang A pp 1105-10.
[7] Rossman L A 2010 Modeling Low Impact Development Alternatives with SWMM *Journal of Water Management Modeling R236-11* pp 167-182.
[8] James W, Rossman L E, and James W R C 2010 User's guide to SWMM 5 *CHI Press Publication R242.*
[9] Decree 80/2014/ND-CP about urban drainage and wase water treatment *Office of Prime Minister of Vietnamese Government*
[10] Technical infrastructure works - Drainage works *National Technical Regulation QCVN 07-2: 2016/BXD.*
[11] Ngoc P 2019 Study on impact of climate change on hydro dynamic regime in Ho Chi Minh City river system *Builders Journal* **7&8** pp 15–35.

Structural performance of reinforced self-compacting concrete columns produced with palm oil fuel ash

K A Mujedu [1,2], M A A-Kadir[1], M Ismail[1,] M N M A Mastor[1], N Zuhan[1], O G Aluko[1,3] and M T M A Sif[1]

[1]School of Civil Engineering, Faculty of Engineering, Universiti Teknologi Malaysia, 81310 UTM Johor Bahru, Johor, Malaysia
[2]Department of Civil Engineering, Faculty of Engineering and Technology, The Federal Polytechnic, Ede, Osun State, Nigeria
[3]Department of Civil Engineering, Faculty of Engineering, Ekiti State University, Ado Ekiti, Nigeria
Email: bayomujedu@yahoo.com; mariyanaida@utm.my; mohammad@utm.my

Abstract. Palm oil fuel ash (POFA) is an agricultural waste produced from the palm oil industries and disposes off into an open ground without any reprocessing or profitable return, has been established to enrich concrete durability as well as mechanical properties when used to replace cement. However the POFA used to partially substitute cement in self-compacting concrete (SCC) production were done on material properties by fabricating standard specimens such as cylinders, prisms and cubes. Moreover, the utilization of POFA to partially substitute cement to cast reinforced concrete structural members is not common compared to fly ash and ground granulated blast furnace slag. This study therefore focuses on structural performance of reinforced self-compacting concrete columns produced with palm oil fuel ash. Two mixes of SCC containing 0% and 15% of POFA were prepared to fabricate the columns. The main variables considered were percentage of POFA content, percentage of steel reinforcement ratio (2% and 3%) and thickness of concrete cover (25 mm and 35 mm). Eight columns were fabricated altogether and tested under axial compression load up to failure. The results of the test revealed that the ultimate axial capacity of reinforced SCC columns produced with POFA is higher than the columns produced without POFA by approximately 2.6% and 2.2% for 25 mm and 35 mm concrete covers, and 2.9% and 2.5% for 2% and 3% steel reinforcement ratios. It is also observed that the load-deflection pattern of reinforced SCC columns produced with and without POFA follow similar trends. However, columns produced with POFA showed slightly lower stiffness than the columns produced without POFA.

1. Introduction

The utilization of concrete is increasing day by day due to the increased in civil infrastructures such as buildings, tunnels, bridges, highways, and so on, and this make the concrete to become the second ever most extensively utilized material in the planet after water. The concrete usage will continue to increase as long as the population of the people continues to increase [1]. At present, the worldwide consumption rate of concrete per year is approximately 30 billion ton [2] compared to the year 2015 which was 25 billion ton [3]. The negative effect of concrete mostly derives from production of cement and this represents its main component. Presently, the worldwide production of cement per year is approximately 4.4 billion tons and this number is predictable to increase to more than 5.5 billion tons by the year 2050, particularly in developing countries [4]. The production processes of cement alone are accountable for approximately 5 – 7% of the global CO_2 anthropogenic emissions and 3% of greenhouse gas global emissions [5]. Also, a higher energy consumption values between 3.2 – 6.3 GJ is used by the cement during the burning process [6]. The release of greenhouse gas and enormous consumption of natural resources associated with cement production cause considerable environmental effect.

Self-compacting concrete (SCC) has numerous merits over the conventional vibrated concrete (CVC), such as noise pollution reduction, removing the necessity for vibration, construction time and cost of labour reduction, enhancing the flowing ability of extremely congested structural elements, reducing the permeability as well as enhancing the concrete durability, easing constructability as well as ensuring better structural performance and enhancing interfacial transitional zone between aggregate and cement paste or steel reinforcement [7]. However, the production of SCC involves the use of chemical admixtures, which are costly. In addition, SCC requires a higher quantity of cement content than CVC and this in turn make the construction cost to be increased. Numerous studies have been executed in order to solve the negative effect associated with cement mentioned above by making use of supplementary cementitious materials (SCMs) to substitute cement in SCC production. The SCMs that have been used are fly ash [8], palm oil fuel ash [3], ground granulated blast furnace slag [9] and rice husk ash [10].

Palm oil fuel ash (POFA) is an agricultural waste produced from the palm oil industries and dispose of into an open grounds without any reprocessing or profitable return [3], has been established to enrich concrete durability as well as mechanical properties when used to replace cement [11,12]. A lot of research investigated on the utilization of POFA to partially substitute cement in SCC production were done on material properties by fabricating standard specimens such as cylinders, prisms and cubes. However, the utilization of POFA to partially substitute cement to cast reinforced concrete structural members is rare in the literatures compared to ground granulated blast furnace slag (GGBFS) [13–15] and fly ash [16–18]. It was only Andalib *et al.* [19] that used combination of POFA and fly ash to cast geo – polymer concrete beams.

Andalib *et al.* [19] carried out an investigation on the structural performance of maintainable waste from palm oil fuel ash and fly ash to produce geo – polymer concrete beams. They reported that the behaviour of reinforced concrete beams produced with POFA – fly ash was almost equal to control reinforced concrete beams because the ultimate moments and cracking patterns of the two types of columns were close together when tested in 90 days. Hawileh *et al.* [14] conducted a research on the performance of reinforced concrete beams cast with 0%, 50%, 70% and 90% of GGBS to replace cement. The results showed that the tensile and compressive strength of the various mixtures were fairly the same. Besides, the performance of RC beams produced with 70% GGBS is the same to the one with 0% GGBS. Lokeshwaran *et al.* [15] carried out a research on the flexural resistance of reinforced self-compacting concrete beams incorporating 0%, 10%, 20%, 30%, 40% and 50% of GGBS to replace cement. They concluded that the flexural behaviour of SCC beams produced with GGBS was improved and exhibits higher ductility than the SCC beams produced without GGBS.

Sunayana and Barai [16] carried out an investigation on the impact of fly ash in using 100% reused aggregate on the flexural performance of reinforced concrete beam. The results showed that the moment carrying capacity at ultimate limit state of beams produced with recycled coarse aggregate and fly ash is mostly similar with the beams produced with natural aggregate. The utilization of fly ash up to 30% and 100% reused coarse aggregate can be used to produce a concrete beam for potential application without compromising its performance in flexure. Sunayana and Barai [17] studied the impact of fly ash in using 100% reused aggregate on the axial compression of reinforced concrete column. They concluded that the results of load carring capacity of columns cast with recycled coarse aggregate incorporating 20% fly ash and 30% fly ash are higher by about 4% and 9% respectively when compared with the columns cast with natural aggregate. Hashmi *et al.* [18] carried out a research on the effect of using 0%, 25%, 40% and 60% fly ash to replace cement on flexural performance of reinforced concrete slab and beam. The results showed that the ultimate and yielding state of reinforced concrete with high volume of fly ash exhibit the same trend when compared to that of reinfoced concrete without fly ash.

All the research carried out on the reinforced strutural members are mostly on beams and only limited data are available on columns. At the same time, the majority of the researchers cast the structural members with the use of CVC while there is limited data on the structural members cast with SCC. Likewise, the 30% of the POFA used to produce geo – polymer concrete beams are mixed with 70% of fly ash. Up till now, the use of POFA to produce SCC structural members are scarce in the literatures to the best knowledge of the authors. To justify the use of POFA in the production of reinforced SCC structural members, more research should be carried out. In this study, an experimental investigation was performed on the structural performance of reinforced SCC columns produced with POFA. The parameters considered were the percentage of POFA (0% and 15%), thickness of concrete cover (25 mm and 35 mm) and percentage of steel reinforcement (2% and 3%). After curing for 28 days, axial loads were applied to all the columns up to failure. The results obtained are discussed in terms of failure patterns, ultimate load capacity, load- deflections and stiffness.

2. Experimental Methods

2.1 Materials

The materials utilized in the present study were dry river sand (fine aggregate), crushed granite (coarse aggregate), POFA, ordinary Portland cement (OPC), tap water and Super plasticizer (SP) of brand name Sika ViscoCrete – 1600. The POFA utilized was procured from the factory (Alif Palm Oil Mill) sited in Kota Tinggi, city of Johor in Malaysia. Details procedure used to process the POFA can be found in Mujedu *et al.* [20]. Table 1 displays the oxide compositions of OPC and POFA while Table 2 displays the physical properties of crushed granite and river sand. The POFA used was confirmed to be class – C type since the combined silica, alumina and iron oxide content is 53% which is above 50% according to ASTM C618 [21].

Table 1. Oxide compositions of OPC and POFA

Compound (%)	SiO_2	Fe_2O_3	Al_2O_3	MgO	CaO	K_2O	MnO	SO_3
OPC	16.2	2.91	3.52	0.76	70.9	0.57	0.13	3.36
POFA	43.9	4.79	4.32	3.1	8.66	10.8	0.06	1.04

Table 2. Physical properties of crushed granite and river sand

Types of aggregates	D_{max} (mm)	FM	Specific gravity
Crushed granite	10	6.3	2.56
River sand	4.75	2.98	2.55

D_{max} : maximum aggregate FM : modulus of fineness

2.2 Mix proportioning of SCC

The proportioning of different concrete constituents was prepared according to EFNARC guidelines and specifications for SCC [22]. Two different SCC mixes were produced with and without the POFA. The first mixes were prepared with POFA at substitution level of 15% by the cement weight while the second mixes were prepared with only OPC. 0.5 water cement – binder ratio was used to prepare all the two mixes. The details summary of the mix proportioning is shown in Table 3. To get the fresh properties in the SCC mixes, Super plasticizer (SP) of brand name Sika ViscoCrete – 1600 was utilized.

Table 3. Mix proportions of SCC containing POFA.

Mix Notation	POFA (% of B)	W/B Ratio	Coarse aggregate (kg/m^3)	Fine aggregate (kg/m^3)	Cement (kg/m^3)	POFA (kg/m^3)	Water (kg/m^3)	SP (kg/m^3)
SCC0	0	0.5	615	750	380	0	190	6.08
SCC15	15	0.5	615	750	323	57	190	6.84

2.3 Design and casting of column specimens

Short columns of 150 mm x 150 mm square in cross-section and 600 mm height were designed as per Eurocode 2 [23]. All the eight columns had 12 mm diameter rebar as longitudinal steel reinforcement and 6 mm stirrup, spacing at 100 mm centre to centre, as transverse reinforcement. The longitudinal steel reinforcement as well as stirrups had yield strength of 538 MPa and 574 MPa respectively. The parameters considered for all the columns are percentage of POFA content of 0% and 15%, concrete covers of 25 mm and 35 mm and longitudinal steel reinforcement ratios of 2.0% and 3.0%. The cross-section of column and reinforcement provided are illustrated in Figure 1. The stirrups provided were bent at 135° with a length of 50 mm to the concrete core as illustrated in Figure 1.

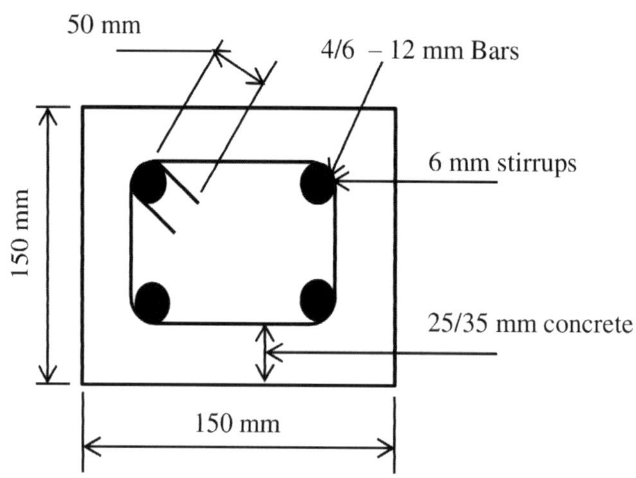

Figure 1. Column cross-section showing design details.

The steel reinforcement cage already prepared was placed inside the steel formwork. Two batches of concrete containing 0% and 15% of POFA content were used for casting all the eight columns. However, the columns were cast one by one from the concrete batch while the casting was done horizontally. Three 100 x 100 x 100 mm cubes were also cast from the concrete batch to determine the compressive strength of concrete. In casting of the columns, the already prepared fresh SCC was poured into the column steel formwork in one layer without compaction. The columns were then levelled and finished on the open side of the formwork. After finishing the casting, the columns were later removed from the formwork after 24±4 hours, marked and stored at ambient condition inside the laboratory. The columns cast were cured by wetting with water every day.

2.4 Instrumentation and test setup

Prior to testing on the machine, the bottom and top ends of all the columns were checked properly in order to make sure that they were all smoothed and levelled. Otherwise, electric grinding machine was used to smooth and level both ends. This is important for obtaining full contact between the ends of the column and loading platens of the testing machine. The column test setup and instrumentations is displayed in Figure 2 (a). Two linear variable differential transducers (LVDT) were installed to the mid-height of each column in horizontal and vertical direction. The LVDT measured the lateral and vertical displacement of the columns. Each column was tested to failure by applying concentric axial load gradually using SOP Tinius Olsen type universal testing machine at a rate of 0.5 kN per second as shown in Figure 2 (b). During the testing of the columns, the data generated was automatically recorded, using a data logger.

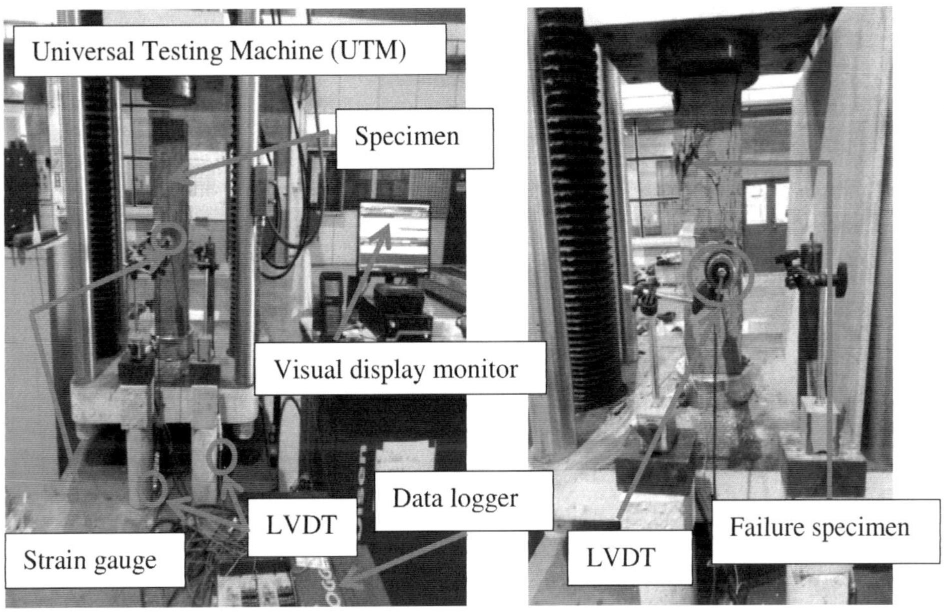

(a) Column test setup (b) Loaded column after failure

Figure 2. Experimental setup

3. Results and discussions

3.1 Compressive strength

Compressive strength test was carried out on SCC cubes cast along with columns as per BS 1881: Part 116 [24]. Compressive strength of SCC cubes produced with 15% POFA (SCC 15) and the one produced without POFA (SCC 0) cured for 28 days is given in Table 4. It can be observed from Table 4 that compressive strength of SCC 15 was 4.9% higher than SCC 0. The increment values of compressive strength noticed for SCC 15 is ascribed to pozzolanic reaction of POFA which reacts with calcium hydroxide that liberated from the cement hydration and thus formed additional calcium silicate hydrate gel which is the major strength provider [11].

Table 4. Compressive strength of test cubes cast with SCC columns produced with and without POFA.

Specimen identification	Compressive strength (MPa)	Average compressive strength (MPa)
SCC 15 - 1	40.75	
SCC 15 - 2	41.01	
SCC 15 - 3	40.13	40.63
SCC 15 - 11	41.13	
SCC 15 - 22	40.08	
SCC 15 - 33	40.68	
SCC 0 - 1	39.26	
SCC 0 - 2	38.82	
SCC 0 - 3	38.17	38.75
SCC 0 - 11	38.69	
SCC 0 - 22	38.42	
SCC 0 - 33	39.14	

3.2 Failure patterns
Failure patterns of reinforced SCC columns after carrying out compression test are discussed under this section. There is longitudinal cracks and crushing of concrete on the surface of all the columns. There is also spalling of concrete cover along the height of the columns. However, the cracks, crushing of concrete and spalling of concrete covers that observed are more noticeable on the columns produced without POFA. Mostly, in all the columns, the failure was brittle, explosive and abrupt.

3.3 Ultimate axial capacity
The ultimate axial capacity of all reinforced SCC columns produced with and without POFA is shown in Table 5. It can be seen from the Table 5 that the ultimate axial capacity of columns produced with POFA is higher than the columns produced without POFA by approximately 2.6%, 2.9%, 2.5% and 2.2%, respectively, for 25 mm and 35 mm concrete covers, and 2% and 3% percentage of steel reinforcements. PCOL1, PCOL2, PCOL3 and PCOL4 stand for columns produced with POFA while CCOL1, CCOL2, CCOL3 and CCOL4 stand for columns produced without POFA. The higher values of ultimate axial capacity observed for columns produced with POFA is ascribed to the pozzolanic reaction of POFA which reacts with the portlandite that released from hydration of cement, and liberated extra calcium silicate hydrate gel which is the major strength provider [11].

The ultimate axial capacity of the columns increased by approximately 12.4%, 11.8% and 12.6%, 12.6% when the percentage of steel reinforcement ratio changed from 2% to 3% for columns produced with and without POFA for 25 mm and 35 mm concrete covers. The increase in ultimate axial capacity of the columns when percentage of steel reinforcement ratio changed from 2% to 3% could be attributed to the large cross-sectional area covered by 3% steel reinforcement ratio. Thus, allow the compressive load coming on to it to spread well compared to 2% steel reinforcement ratio which has small cross-sectional area to contain the spread load.

When the concrete cover changed from 25 mm to 35 mm, the ultimate axial capacity of the columns increased by approximately 1.9%, 1.3% and 1.6%, 1.6% for columns produced with and without POFA for 2% and 3% steel reinforcement ratios. The increase in ultimate axial capacity of the columns when concrete cover changed from 25 mm to 35 mm could be attributed to the reduction in the buckling of steel reinforcements.

Table 5. Test results of reinforced SCC columns produced with and without POFA.

Column No	% of steel reinforcement	Concrete cover (mm)	Ultimate axial capacity (kN)	Secant stiffness (kN/mm)
PCOL 1	2	25	835.1	355.4
CCOL 1	2	25	813.6	360.0
PCOL 2	2	35	850.6	322.2
CCOL 2	2	35	826.6	324.2
PCOL 3	3	25	938.6	430.6
CCOL 3	3	25	916.1	436.2
PCOL 4	3	35	951.1	393.0
CCOL 4	3	35	930.6	404.6

3.4 Load-deflection pattern

The relationship of applied load with axial displacement of the columns produced with POFA and without POFA is shown in Figure 3. It can be seen from the Figure 3 that when the same load level was applied to the columns, the columns produced with POFA showed slightly smaller rigidity than the columns produced without POFA and that is why the values of axial displacement displayed by the columns produced with POFA were slightly higher than the columns produced without POFA. The slightly higher values of axial displacement displayed by the columns produced with POFA are attributed to its self-weight which is slightly less than the columns produced without POFA.

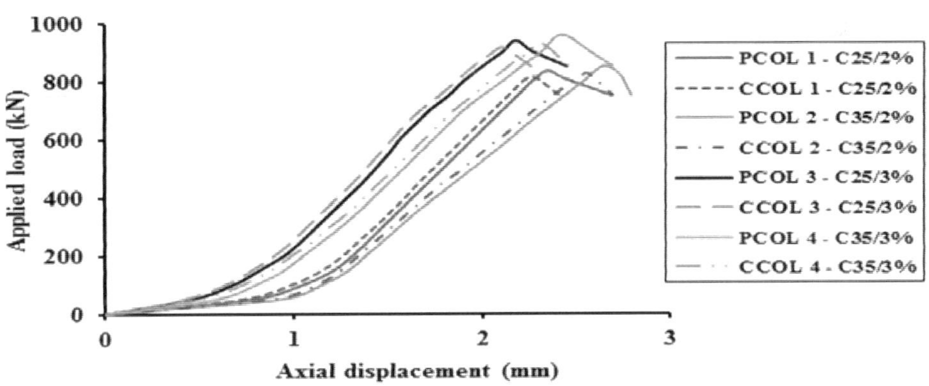

Figure 3. Applied load – axial displacement relationship of reinforced SCC columns produced with and without POFA

3.5 Stiffness

The stiffness of the columns produced with and without POFA is illustrated in Table 5. The secant stiffness of all the columns produced was determined by dividing the ultimate axial capacity by the maximum axial displacement recorded during the testing. Though the value may not signify the total actual stiffness of all the columns produced [25]. However, it offers a valuable criterion to compare the stiffness of columns produced with and without POFA. It can be observed from the Table 5 that the stiffness of columns produced with POFA is slightly lower than the one produced without POFA.

This is accredited to its self-weight which is slightly less than the columns produced without POFA. For the two types of column produced, the stiffness is reduced as the concrete cover increased. However, the stiffness is increased as the percentage of steel reinforcement increased.

4. Conclusions

The structural performance of reinforced self-compacting concrete columns produced with 15% POFA was investigated. The findings from the experiment carried out are as follows:

- The compressive strength of cubes produced with POFA is higher than the cubes produced without POFA by approximately 4.9%.
- The cracks, crushing of concrete and spalling of concrete covers that observed are more noticeable on the columns produced without POFA.
- The ultimate axial capacity of columns produced with POFA is higher than the columns produced without POFA by approximately 2.6%, 2.9%, 2.5% and 2.2%, respectively, for 25 mm and 35 mm concrete covers, and 2% and 3% percentage of steel reinforcements.
- The ultimate axial capacity of columns increased by approximately 12.4%, 11.8% and 12.6%, 12.6% when the percentage of steel reinforcement ratio changed from 2% to 3% for columns produced with and without POFA for 25 mm and 35 mm concrete covers.
- When the concrete cover changed from 25 mm to 35 mm, the ultimate axial capacity of columns increased by approximately 1.9%, 1.3% and 1.6%, 1.6% for columns produced with and without POFA for 2% and 3% steel reinforcement ratios.
- The stiffness of columns produced with POFA is slightly lower than the one produced without POFA.

Acknowledgments

The authors honestly appreciate the support provided by the Research Management Centre (RMC) through Grant No: 16H94 and the Technical staff of Structure and Materials unit, Universiti Teknologi Malaysia (UTM). The financial supports given by the Federal Government of Nigeria through TETFund are thankfully appreciated by the first author. Many thanks are also extending to authority of Alif Palm Oil Mill, Malaysia for providing the palm oil fuel ash.

References

[1] Henry M and Kato Y 2014 Understanding the regional context of sustainable concrete in Asia : Case studies in Mongolia and Singapore. *Resources, Conserv. Recycl.* **82** 86–93.
[2] Joshi S, Goyal S and Reddy M S 2020 Influence of biogenic treatment in improving the durability properties of waste amended concrete : A review. *Constr. Build. Mater.* **263** 120170.
[3] Alsubari B, Shafigh P and Jumaat M Z 2016 Utilization of high-volume treated palm oil fuel ash to produce sustainable self-compacting concrete. *J. Clean. Prod.* **137** 982–96.
[4] Merli R, Preziosi M, Acampora A, Lucchetti M C and Petrucci E 2020 Recycled fibers in reinforced concrete : A systematic literature review. *J. Clean. Prod.* **248** 119207.
[5] Heede P V D and Belie N De 2012 Environmental impact and life cycle assessment (LCA) of traditional and "green" concretes : Literature review and theoretical calculations. *Cem. Concr. Compos.* **34** 431–42.
[6] Van Oss H G and Padovani A C 2003 Cement Manufacture and the Environment. Part II: Enviromental Challenges and Opportunities. *J. Ind. Ecol.* **7** 93–126.
[7] Shi C, Wu Z, Lv K and Wu L 2015 A review on mixture design methods for self-compacting concrete. *Constr. Build. Mater.* **84** 387–98.
[8] Kristiawan S A and Aditya M T M 2015 Effect of high volume fly ash on shrinkage of self-compacting concrete. *Procedia Eng.* **125** 705–12.
[9] Altoubat S, Badran D, Junaid M T and Leblouba M 2016 Restrained shrinkage behavior of

Self-Compacting Concrete containing ground-granulated blast-furnace slag. *Constr. Build. Mater.* **129** 98–105.

[10] Raisi E M, Amiri J V and Davoodi M R 2018 Influence of rice husk ash on the fracture characteristics and brittleness of self-compacting concrete. *Eng. Fract. Mech.* **199** 595–608.

[11] Ranjbar N, Behnia A, Alsubari B, Birgani P M and Jumaat M Z 2016 Durability and mechanical properties of self-compacting concrete incorporating palm oil fuel ash. *J. Clean. Prod.* **112** 723–30.

[12] Alsubari B, Shafigh P, Ibrahim Z, Alnahhal M F and Jumaat M Z 2018 Properties of eco-friendly self-compacting concrete containing modified treated palm oil fuel ash. *Constr. Build. Mater.* **158** 742–54.

[13] Sangeetha S P and Joanna P S 2014 Flexural Behaviour of Reinforced Concrete Beams with GGBS. *Int. J. Civ. Eng.Technol.* **5** 124–31.

[14] Hawileh R A, Abdalla J A, Fardmanesh F, Shahsana P and Khalili A 2017 Performance of reinforced concrete beams cast with different percentages of GGBS replacement to cement. *Arch. Civ. Mech. Eng.* **17** 511–19.

[15] Lokeshwaran N, Bhuvaneshwari M, Mirunalini V and Sakthivel M 2017 Flexural Resistance of Self Compacted Concrete Beams by Partial Replacement of OPC with GGBS. *Int. J. Civ. Eng. Technol.* **8** 479–93.

[16] Sunayana S and Barai S V 2018 Flexural performance and tension-stiffening evaluation of reinforced concrete beam incorporating recycled aggregate and fly ash. *Constr. Build. Mater.* **174** 210–23.

[17] Sunayana S and Barai S V 2019 Performance of fly ash incorporated recycled aggregates concrete column under axial compression : Experimental and numerical study. *Eng. Struct.* **196** 109258.

[18] Hashmi A F, Shariq M and Baqi A 2020 Flexural performance of high volume fly ash reinforced concrete beams and slabs. *Structures* **25** 868–80.

[19] Andalib R, Hussin M W, Abd Majid M Z, Azrin M and Ismail H H 2014 Structural Performance of Sustainable Waste Palm Oil Fuel Ash - Fly Ash Geo - Polymer Concrete Beams. *J. Environ. Treat. Tech.* **2** 115–19.

[20] Mujedu K A, Ab-kadir M A and Ismail M 2020. Effect of high temperatures on physical and compressive strength properties of self-compacting concrete incorporating palm oil fuel ash. *IOP Conf. Ser. Mater. Sci. Eng.* **849** 1–11.

[21] ASTM C618 2019 Standard specification for coal fly ash and raw of calcined natural pozzolan for use in concrete. *ASTM International*, West Conshohocken, PA.

[22] EFNARC S 2005 The European Guidelines for Self - Compacting Concrete, Specification, Production and Use.

[23] EN 1992-1-1. Eurocode 2 2004 Design of concrete structures - Part 1-1: General rules and rules for buildings, Brussels.

[24] BS 1881 1983 Testing Concrete - Part 116: Method for determination of compressive strength of concrete cubes. *British Standards Institution.*

[25] Al-Gasham T S, Mhalhal J M and Jabir H A 2019 Influence of post-heating on the behavior of reinforced self-compacting concrete hollow columns. *Structures* **22** 266–77.

IOP Conf. Series: Materials Science and Engineering 1153 (2021) 012007 doi:10.1088/1757-899X/1153/1/012007

Stability of Railway Embankment in Saturated and Unsaturated Conditions

M J Roshan[1,3*], A S A. Rashid[1], N A Wahab[1], M A Hezmi[1], S N Jusoh[1], M Azmi[2]

[1] School of Civil Engineering, Universiti Teknologi Malaysia, Johor Bahru, Malaysia
[2]School of Civil Engineering, Universiti Sains Malaysia, Pulau Pinang Malaysia
[3] Department of Railway Engineering, Kabul Polytechnic University, Afghanistan
*Corresponding author: mj.roshan@kpu.edu.af

Abstract- Slope failure is a vital issue in civil engineering infrastructures. Besides, climate change implications made this problem more critical. Therefore, in this study, a railway embankment's slope stability has been studied in saturated and unsaturated conditions. The soil sample was collected from Universiti Teknologi Malaysia, Johor Bahru. The necessary and required data obtained through lab experiments. The obtained data from lab experiments then used in Seep/W and Slope/W to model the railway embankment numerically. The results obtained from this study indicate that rainfall infiltration plays a vital role in the safety factor value. The obtained results also depicted that both rainfall intensity and duration play an important role in slope stability. The factor of safety showed a declining pattern with increasing rainfall and duration of rainfall. The safety factor decrement occurred since the matric suction was reduced (negative pore water pressure) by rainfall infiltration. Besides, the safety factor decrease was more acute when train load was applied in a rainy period. Thus, it is suggested that in a long period of rainy time, the railway authorities should stop the train operation, or the embankment stability should be increased using some stabilizations methods in the planning and designing stage.

1. Introduction

Climate change implications can be seen in varying ways. One of these implications is the changing rainfall pattern, one of the most critical factors in slope instability. Generally, in all parts of the world and particularly in tropical regions, rainfall-induced landslides and slope failure are the most common [1, 2]. The sustainability in urban living may be affected by slope instability caused by rainfall [2]. Although most natural and human-made slopes are in unsaturated conditions, the influence of matric suction on slope stability analysis is mainly ignored in studies. The vanishing of negative pore water pressure with increasing moisture content and infiltrated rainfall is counted as one of the main reasons for matric suction to be ignored in slope stability analysis. However, the matric suction can be maintained by adopting some measures such as slope surface cover and surface compaction to minimize the infiltrated rainfall water [3].

The railway embankment is typically in unsaturated condition because of its height from the ground surface and deep water table. However, during the rainy seasons, the embankment soil condition changes from unsaturated to saturated condition. The location of the water table may be shifted from a down position toward the ground surface. In the rainy season, the prevention of entering water into the ballasted railway track layers is roughly impossible [4]. The embankment, therefore, experiences cyclic wetting and drying condition during its service life. Thus, this cyclic change of soil conditions results in decreased shear strength. The shear strength along a potential failure surface in a slope can be reduced with infiltrated rainfall water [5]. In this regard, Zi-Zhen et al. [6] stated that decreased soil cohesion plays the leading role in most short-term slope failures. The slope failure may be happened either by rising phreatic level or directly by infiltrated rainfall [7].

In regard to the rainfall impacts, numerous researches have been conducted on slope stability worldwide [8-12]. The influence of rainfall on slope stability has been studied through field instrumentation [9], laboratory-based physical modelling [10,11], and numerical modelling [12]. Having used Singapore climate data, Rahardjo et al. [13] studied the effect of flux boundary conditions on pore water pressure variation in a slope. In contrast, Yunusa et al. [14] studied it with consideration of Malaysia climate. Zhang et al. [3] performed research on the effect of rainfall infiltration in a slope using Seep/w both for the steady-state and transient condition. The studied slope inclined 30° with a horizontal surface, and the soil has been considered isotropic and homogenous. The obtained results indicate that the matric suction is mainly affected by ground surface moisture flux. In transient condition rainfall flux, the saturated coefficient of permeability and water storage function showed a significant role in pore water pressure variation.

As mentioned earlier, many researchers have studied the influence of infiltration on slope stability, but most studies have been carried out on the natural slope. Thereby, it can be mentioned that the research studies on the rainfall-induced railway embankment slope instability are not sufficient in the available literature. However, some researches on railway embankment stability can be found in the literature [18-25]. Despite that, most of the available studies in the literature neglected the coupling effects of train load and rainfall water infiltration on the slope stability analysis. Therefore, the aim of this paper is to study the effect of rainfall infiltration and applied trainload on safety factors using Seep/W and Slope/W.

2. Theory

2.1. *Water flow*

The water flow calculation in the saturated soil is based on Darcy's law, according to Equation 1.[3].

$$v = -k\frac{\partial h}{\partial y} \qquad (1)$$

V is equivalent to flow rate or flux, illustrating the Darcian velocity, k is hydraulic conductivity, and $\partial h/\partial y$ is hydraulic gradient. Equation 1 is based on constant hydraulic conductivity, which is one of the saturated soil features. Equation 1 can also be used for water velocity in the unsaturated condition, provided that the k should be considered as a function of matric suction. The flow of water in the soil for saturated and unsaturated conditions is different [16]. Equation 2 can be utilized for a transient flow of water through unsaturated soil [17].

$$\frac{\partial}{\partial x}(k_x\frac{\partial h}{\partial x}) + \frac{\partial}{\partial y}(k_y\frac{\partial h}{\partial y}) + Q = \frac{\partial \theta}{\partial t} \qquad (2)$$

Where h is the total hydraulic head, k is hydraulic conductivity, Q is the applied boundary flux, θ, is the volumetric water content and, t is time.

2.2. *Factor of safety*

In civil engineering projects analyzing of stability of slope is of paramount importance. For doing so, different methods such as the finite element method and limit equilibrium method can be utilized. In this research, the Morgenstern-Price method, which is based on the limit equilibrium theory, is used to calculate the factor of safety. The moment equilibrium and force equilibrium are two factors that the limit equilibrium theory is based on [18]. Therefore in the saturated condition, Equation 3 and Equation 4 can be used for force and moment equilibrium safety factor, respectively [19]:

$$F_f = \frac{\sum[c'\beta\cos\alpha + \{N - U_w\beta\}\tan\phi'\cos\alpha]}{\sum N\sin\alpha} \tag{3}$$

$$F_m = \frac{\sum[c'\beta R + \{N - U_w\beta\}R\tan\phi']}{\sum Wx - \sum Nf} \tag{4}$$

Where, c', is the effective stress, β, is the slope length at the bottom of the slice, α, is the inclination of slice base, N, normal force at the bottom of the slice, U_w, is pore water pressure, ϕ', effective friction angle, W, is total slice weight, and R, x, f, are geometric parameters.

In the same way, to consider the matric suction effects on shear strength of soil, Equation 5 and Equation 6 are used for the safety factor calculations based on force and moment, respectively[19].

$$F_f = \frac{\sum[c'\beta\cos\alpha + \{N - U_w\beta\frac{\tan\phi^b}{\tan\phi'}\}\tan\phi'\cos\alpha]}{\sum N\sin\alpha} \tag{5}$$

$$F_m = \frac{\sum[c'\beta R + \{N - U_w\beta\frac{\tan\phi^b}{\tan\phi'}\}R\tan\phi']}{\sum Wx - \sum Nf} \tag{6}$$

Where, ϕ^b, is the angle that represents the share of matric suction in the shear strength.

3. Method and Materials

3.1. Laboratory experiments

In the current study, the soil used as embankment material was collected from Universiti Teknologi Malaysia, Johor Bahru. The coordinate of the selected site for soil sampling is 1°33'32.9"N 103°38'39.4"E. The collected samples were then tested in the laboratory. Based on the USCS system [20], the soil is classified as elastic silt or silty clay(MH), while based on the British Soil Classification System (BSCS), it is categorized as high plasticity sandy silt (MHS). Table 1 illustrates some basics and mechanical characteristics of subgrade, while Table 2 summarizes the used material's hydraulic features. The soil-water characteristic curve of subgrade was obtained using the pressure plate extractor (i.e., axis translation) method [21]. However, the lab experiment data is a few obtained result points that cannot be used effectively for numerical modelling. In other words, the continuous soil water characteristics curve (SWCC) in the desired matric suction range is of importance in numerical modelling [22]. Thus, the fitting parameters were then generated using the SWRC-Fit software [36][37]. The result obtained from the pressure plate test used in SWRC-Fit software and the acquired graph illustrates good compatibility of test results with Van Genuchten (1980), Fredlund and Xing (1994), and Brooks and Corey (1964) fitting models, as shown in Figure 1. It should be noted that since the application of drying SWCC yields a lower factor of safety than wetting and combined SWCC [23], thus in this study, the considered SWCC is a drying pattern. Furthermore, the saturated hydraulic conductivity of the soil has been obtained using the falling head method. The unsaturated hydraulic conductivity was calculated with the indirect measurement method, using the SWCC and saturated hydraulic conductivity, as shown in Figure 2.

IOP Conf. Series: Materials Science and Engineering 1153 (2021) 012007 doi:10.1088/1757-899X/1153/1/012007

Table 1. Index and mechanical characteristics of subgrade

Properties	Symbol	Value
Liquid limit: %	l_l	70.3
Plastic limit: %	Pl	42
Plasticity index: %	Pi	28.3
Gravel: %	-	12.79
Sand: %	-	17.54
Silt: %	-	61.26
Clay: %	-	8.41
Soil classification:	-	
USCS		MH
AASSHTO		A-7-5
Optimum moisture content: %	OMC	28
Maximum dry density: g / cm^3	MDD	1.39
Specific gravity	G_s	2.74
Effective cohesion: KPa (Based on previous studies)	c'	7.2
Effective friction angle: ° (Based on previous studies)	ϕ'	32
Friction angle related to matric suction:	ϕ^b	$\phi^b = 2/3\phi'$
Dry unit weight: KN / m^3	γ	13.64
Poison ratio	ν	0.35

Table 2. Subgrade hydraulic characteristics

Properties	Value
Saturated hydraulic conductivity Ks, (m/s)	1.9×10^{-8}
n	0.78
m	1.87
a	189.41
Residual volumetric water content, θ_r	0.23
Saturated volumetric water content, θ_s	0.54

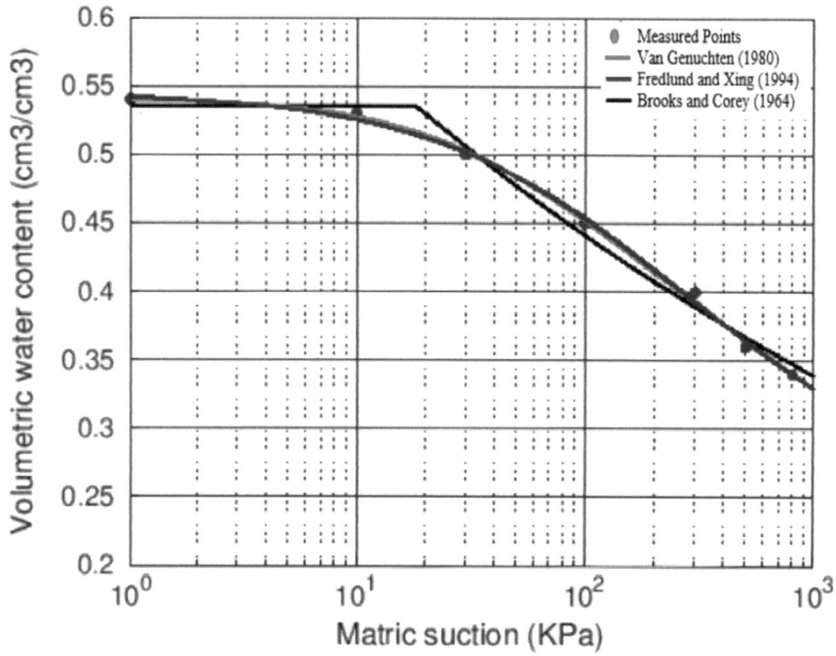

Figure 1. Subgrade SWCC

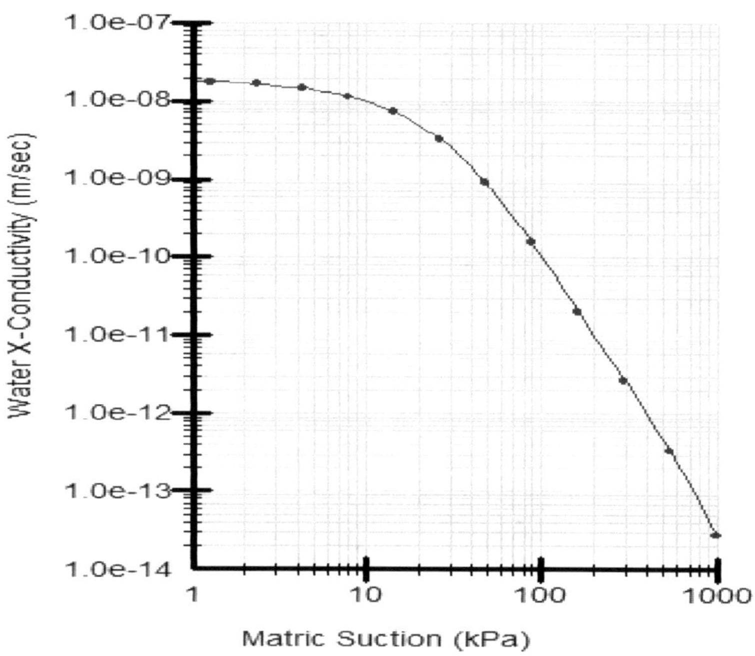

Figure 2. Unsaturated hydraulic conductivity

The ϕ^b factor considering the influence of matric suction $(u_a - u_w)$ in shear strength of soil takes various values according to the degree of saturation. For instance, in the saturated condition, the

$\phi^b = \phi'$ [24] but for the unsaturated condition, it can take a value of $2/3\phi'$ [14]. Therefore, in the current study, for unsaturated condition ϕ^b value is considered as $2/3\phi'$.

3.2. Rainfall Data

The rainfall data is the paramount importance in slope stability analysis to be considered. This issue's importance is doubled for tropical regions where the annual rainfall intensity is considerably high [10]. The frequent rainfall has caused many slope failures in tropical regions, including Malaysia [8]. The ponding does not occur when the rainfall intensity is low(i.e., the rapid ponding happens when the rainfall intensity is greater than 10mm/h) [6]. In other words, all the rainfall water can infiltrate if the saturated permeability is equal to or greater than the infiltration rate [25]. In this regard, Kassim et al. [26] conducted detailed research simulating the effect of different permeability values on negative pore water pressure. The obtained results depicted the ponding phenomena when the infiltration rate was more significant than saturated hydraulic conductivity ($q > K_s$) of the top layer (silty sand).

In the current research, the rainfall data obtained from the Department of Irrigation and Drainage of Malaysia for 2019 and 2020. The rainfall data used in this study have been recorded in the Senai metrology station, located in University Technology Malaysia. Figure 3 depicts the monthly rainfall of Johor Bahru based on the Senai metrology station. In Figure 3, it is seen that December has the maximum value rainfall of 403.5mm, while July month possesses 29.5 mm, which is the minimum value. Therefore, December and July months are selected in this research that Figure 4 depicts daily rainfall intensity for selected months. The whole throughfall does not percolate to the ground. Thus, the rainfall infiltration rate has been taken differently by researchers. For instance, Ng et al. [27] have considered 60% and 40% precipitation for infiltration and run-off, respectively, while Rahardjo et al.[28] have suggested 40-74% of total rainfall amount as infiltration value. However, in this study, 70% and 30% of rainfall are taken as infiltration and run-off value, respectively [29, 14].

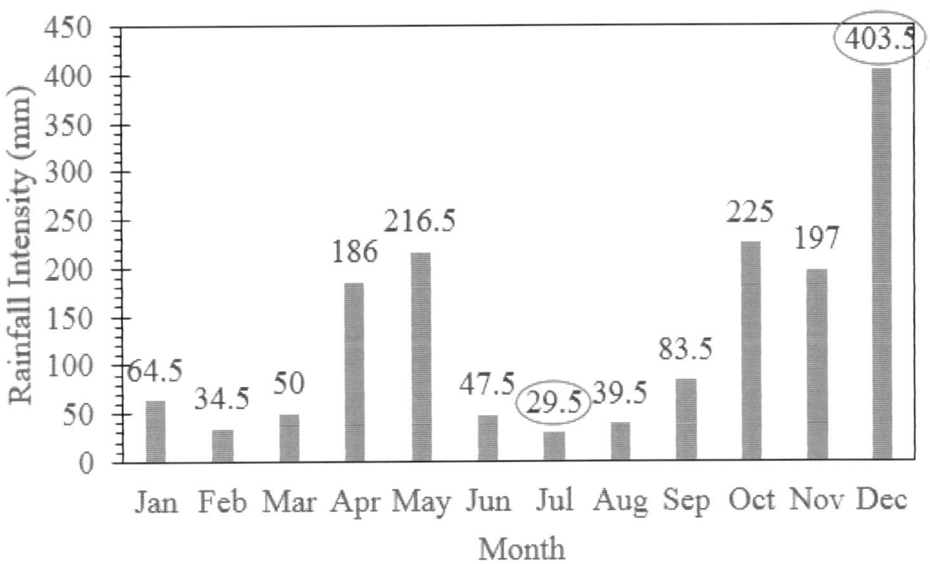

Figure 3. Monthly rainfall

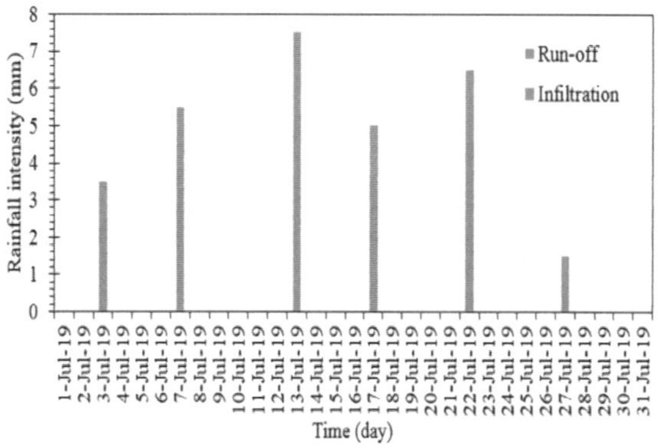

(a)

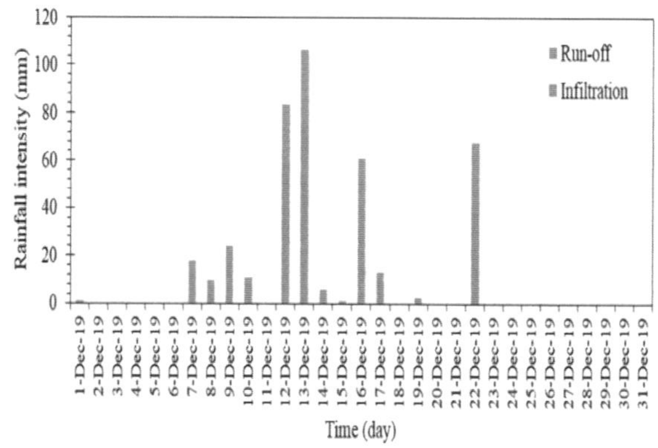

(b)

Figure 4. Daily rainfall intensity a) December 2019 b) July 2019

3.3. *Numerical Modelling*

Slopes can be categorized into two types, namely natural slope and human-made slope. Factors causing the slope failure can be varying, but most of the slope failure (i.e., 60%) caused by design errors [30]. Selection of unsuitable slope inclination and high (i.e., slope geometry) and incapability in finding the load and soil strength parameters are the most common design errors causing slope instability [31]. Therefore, the railway embankment slope must be planned and designed carefully and effectively in the early planning stages. In this study, the slope of the embankment is considered 1.5H:1V. For avoiding the dimension effects on numerical modelling, the dimensions of the geometry model are taken with respect to [16]. The applied load is considered for the maximum speed of 160km/h and axle load of 18.5ton. The details of track and train are illustrated in Table 3.

Table 3. Specifications of track and train

Parameters	Value
Gauge	1435 mm
Axle load	18.5 ton (184.33kN)
Speed	160 km/h
Dynamic impact factor	1.54
Ballast thickness	0.4m
Sleeper:	
Length	2.6 m
Width	0.3 m
Height	0.224 m

For calculating the tangent pressure in the interface of sleeper and ballast, the applied load on rail and sleeper interface needs to be calculated. For doing so, Equation 7 can be used [32]. Afterward, Equation 8 is utilized to calculate the impact factor due to speed. Finally, Equation 9 is applied to compute the value of tangential stress on sleeper and ballast interface [33].

$$q_r = 0.5IF \times P \quad (7)$$

$$IF = 1 + \frac{4.5V^2}{10^5} - \frac{1.5V^3}{10^7} \quad (8)$$

$$P_a = \frac{2q_r}{A} \tag{9}$$

Where q_r (kN) is applied load on the interface of rail and sleeper, P (kN) represents the wheel load, IF is the impact factor, V (km/h), is design speed, A is the contact area of sleeper with ballast, and P_a (kPa), is the tangential stress on the interface of sleeper and ballast. Having considered the specifications of Table1, the P_a =181.96 kPa is obtained using Equation 9. Since the ballast layer is not considered in this modelling, the maximum applied stress on the surface of subgrade needs to be found due to 0.4m ballast thickness using Equation 10.

$$P_c = \frac{0.17P_a}{h^{1.25}} \tag{10}$$

P (kPa), represents the maximum stress on the top of the subgrade, while h (m) shows the thickness of granular material on the top of the subgrade. Thus, the applied maximum stress on the top of the subgrade is acquired 97kPa based on the considered ballast thickness.

The flux boundary condition is considered on the ground surface (i.e., 8-7-6-5-4-3) to model the influence of rainfall on matric suction and safety factor. Besides, the water table's location is assumed 2m below the base of the embankment. Briefly, Figure 5 illustrates the flowchart of numerical modelling, and Figure 6 depicts embankment geometry. Section a-a in Figure 6 is illustrated to evaluate the matric suction change due to rainfall infiltration.

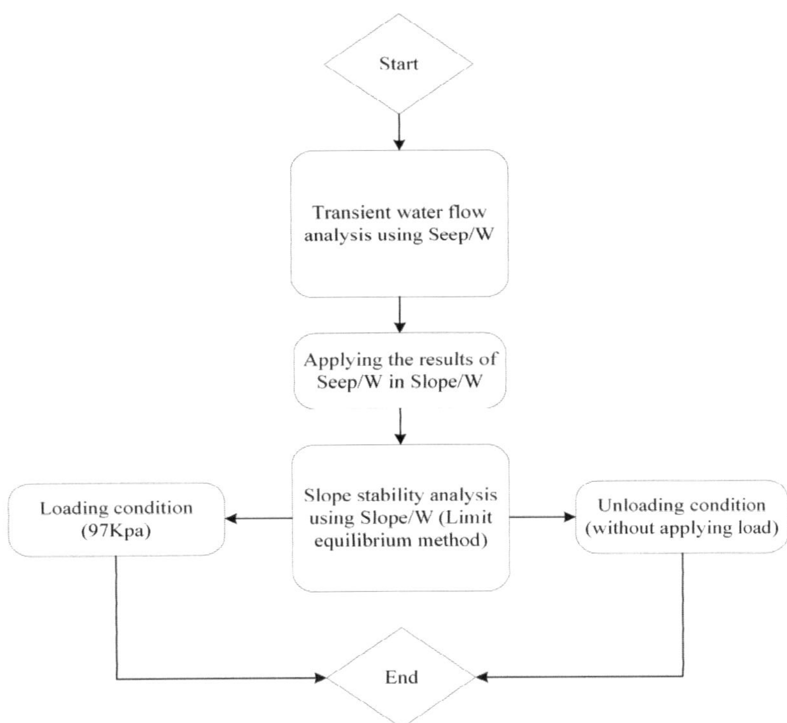

Figure 5. Flowchart of numerical analysis

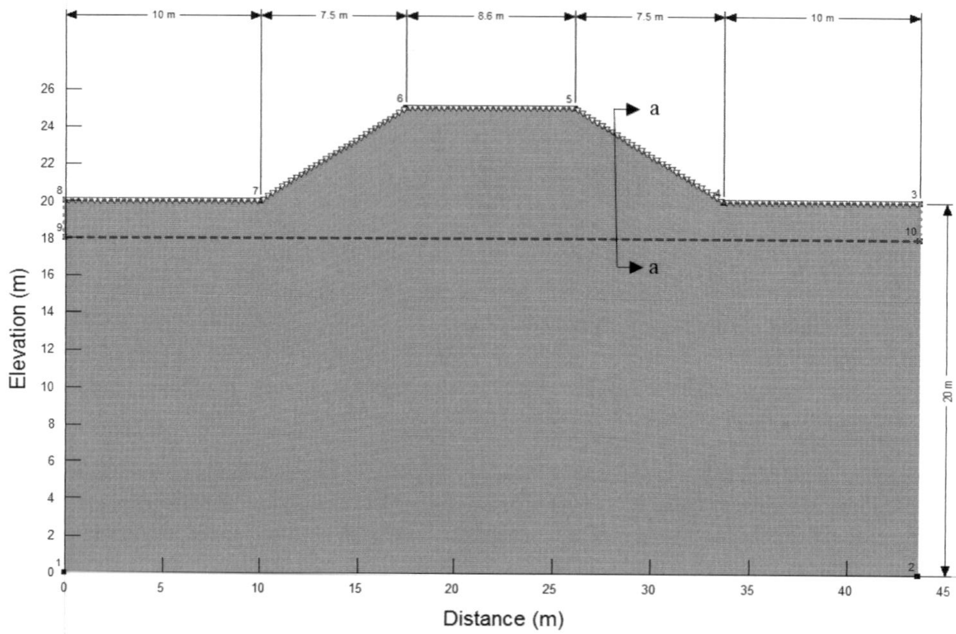

Figure 6. Geometry of embankment

Although different methods are available in Seep/W for slope stability analysis, in this research, the Morgenstern-Price method that considers both moment equilibrium and force equilibrium have been applied in numerical modelling.

4. Results and Discussions

The safety factor decrease can be occurred due to reduced shear strength and increased shear stress. The strength decline can be caused by increasing moisture content and decreasing matric suction, while shear stress is caused by external factors such as applying load and traffic load. The rate of matric suction dissipation relates to rainfall intensity, rainfall duration, and soil permeability [3]. In regard to tropical regions' climate conditions, it can be stated that 80% of slope instability occurring in tropical regions is related to matric suction dissipation [34]. The shear strength reduction during the rainfall infiltration is caused by decreased matric suction along the potential slip surface [22]. This phenomenon can be clearly explained in section a-a of Figure 6 in Figure7. The negative pore-water pressure (Matric suction) exists above the water table before starting the rainfall, but the decreasing matric suction starts with rainfall. Besides, Figure 7b illustrates that during December, the decrement is more than that of July, as depicted in Figure 7a. This issue is attributed to rainfall intensity, which is higher in December compared with July.

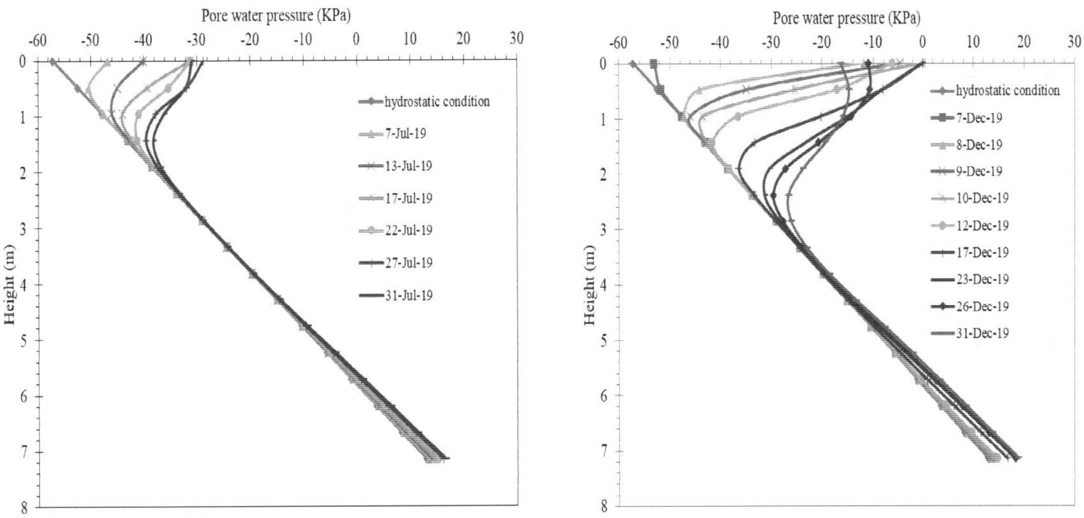

(a) (b)

Figure 7. Pore water pressure in section1-1 a) for July b) for December

Moreover, depth and time are two parameters that govern pore water pressure [35]. As shown in Figure7, at the beginning (hydrostatic condition), the pore water pressure is negative above the water table. Afterward, the matric suction decreases due to rainfall intensity over time. Besides, looking at Figure 7, it can be readily understood that the rate of change in negative pore water pressure due to rainfall is getting smaller with depth. Therefore, as shown in Figure 8, the factor of safety (FS) shows a decreasing pattern when the embankment is subjected to rainfall infiltration. Whereas Figure 9 depicts the influence of both rainfall infiltration and applied load on the safety factor reduction pattern. By comparing Figure 8 and Figure 9, it can be readily perceived that the applied load excessively diminished the factor of safety (FS) when incorporated with rainfall percolation. However, the rainfall alone reduces the safety factor. Nevertheless, its influence is not significant when both load and infiltration have been applied to the embankment together.

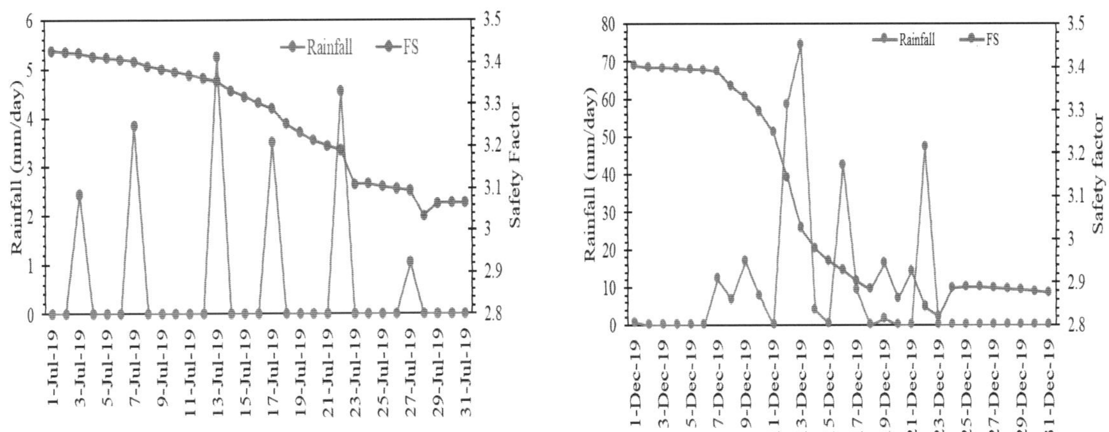

(a) (b)

Figure 8. Influence of rainfall on factor of safety a) in July b) in December

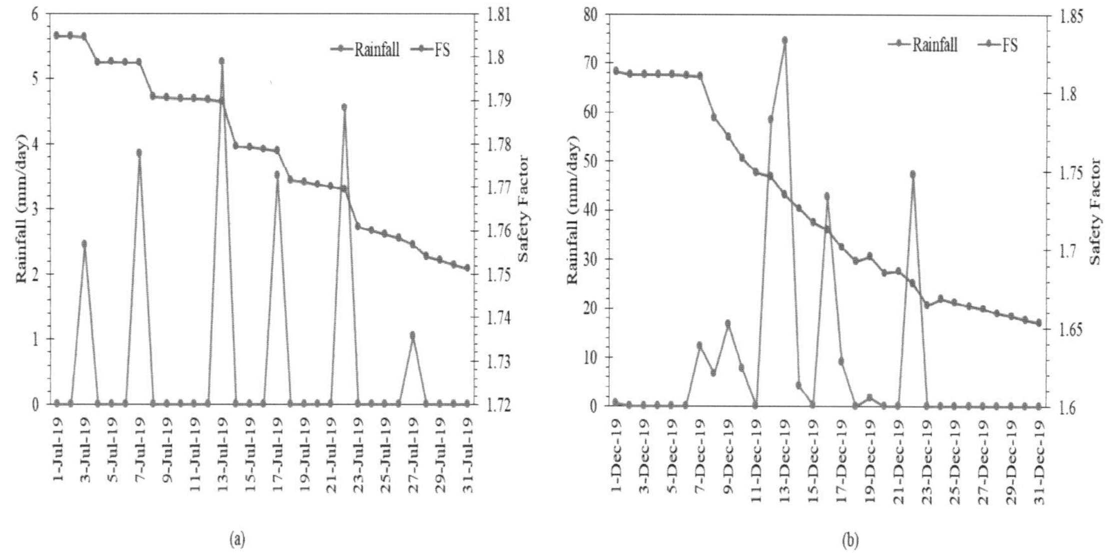

Figure 9. Influence of rainfall and trainload on the factor of safety a) in July b) in December

Succinctly the influence of rainfall on safety factor can be attributed to the decreased shear strength due to positive excessive pore water pressure or decreasing matric suction, as can be seen in Figure 7. On the other hand, the applied load's influence on diminishing the safety factor can be ascribed to the increased shear stress. Figure 10 illustrates the slip surface location at the end of December, where the safety factor is 2.875 and 1.654 for unloading and loading conditions, respectively.

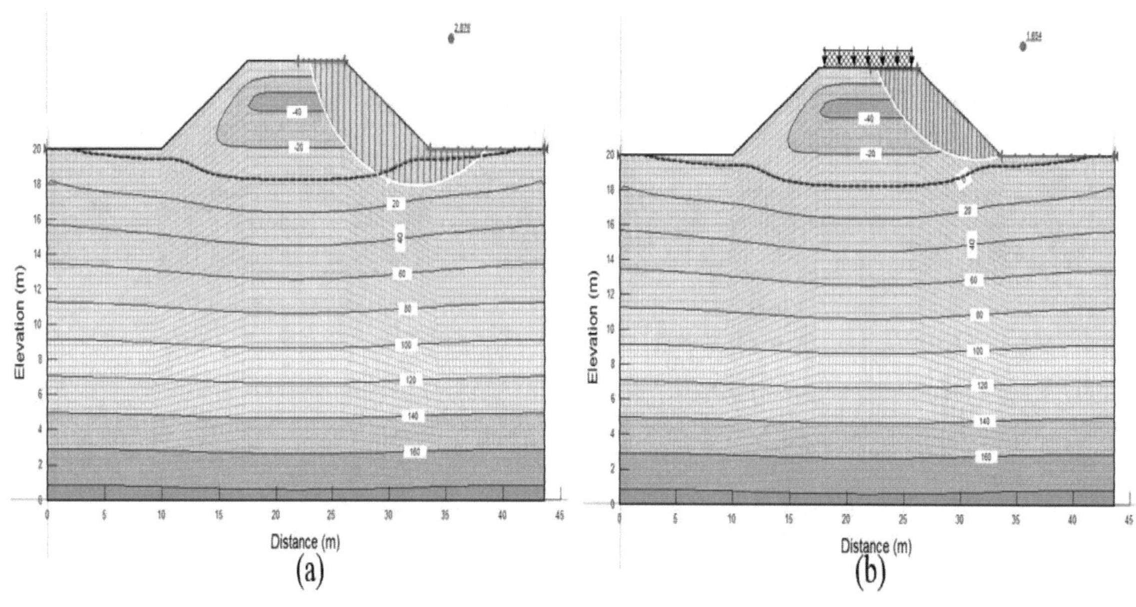

Figure 10. Slip surface at the end of December a) effect of rainfall b) effect of rainfall and load

IOP Conf. Series: Materials Science and Engineering 1153 (2021) 012007 doi:10.1088/1757-899X/1153/1/012007

5. Conclusions

This research has been conducted on the stability analysis of a presumed railway embankment. Based on the results obtained, the following conclusion can be drawn:

- The decreasing water content results in increasing matric suction.
- The permeability in the unsaturated condition is a matric suction function; the permeability decreases with increasing matric suction.
- The increasing positive pore water pressure or decreasing matric suction reduces the safety factor resulting in railway instability.
- The rainy seasons are more critical for railway embankment slope since the positive pore water pressure increases excessively in the rainy seasons.
- The decreasing level of matric suction depends on the depth where rainfall water can infiltrate, and beyond that depth, the matric suction does not change.
- The applied load on the railway track has a negative influence on railway embankment stability, and this influence becomes more intense with train operation during the rainy seasons.

Acknowledgments

The Fundamental Research Grant Scheme supported this work (FRGS) awarded the Ministry of Education of Malaysia (Engineering and Microstructural Characteristics of Lateritic Soil Treated with Ordinary Portland Cement Undercyclic Saturated (Wetting) and Unsaturated (Drying) Conditions - FRGS/1/2019/TK01/UTM/02/13). The first author also appreciates the Ph.D. opportunity provided by the Higher Education Development Program (HEDP) under the Afghanistan Ministry of Higher Education.

References

[1] Kassim A, Gofar N, Lee L M and Rahardjo H 2012 Modeling of suction distributions in an unsaturated heterogeneous residual soil slope *Eng. Geol.* 131–132 70–82

[2] Rahardjo H, Satyanaga A and Leong E C 2016 Effects of rainfall characteristics on the stability of tropical residual soil slope *E3S Web of Conferences 9* vol 15004 pp 1–6

[3] Zhang L L, Fredlund D G, Zhang L M and Tang W H 2004 Numerical study of soil conditions under which matric suction can be maintained *Can. Geotech. J.* 41 569–82

[4] Ghataora G S and Rushton K 2012 Movement of water through ballast and subballast for dual-line railway track *J. Transp. Res. Board* 11 78–86

[5] Gui M W and Wu Y M 2014 Failure of soil under water infiltration condition *Eng. Geol.* 181 124–41

[6] Zi-zhen L, Zhi-xin Y, Zhan-hong Q, Xiao-gang W and Ji-wei L 2020 Stability analysis of an unsaturated soil slope considering rainfall infiltration based on the Green-Ampt model *J. Mt. Sci.* 17 2577–90

[7] Smith P G C 2003 *Numerical analysis of infiltration into partially saturated soil slopes* (Imperial College of Science, Technlogy and Medicine)

[8] Lee L M, Gofar N and Rahardjo H 2009 A simple model for preliminary evaluation of rainfall-induced slope instability *Eng. Geol.* 108 272–85

[9] Rahardjo H, Nio A S, Harnas F R and Leong E C 2014 Comprehensive instrumentation for real time monitoring of flux boundary conditions in flope *Procedia Earth Planet. Sci.* **9** 23–43

[10] Lee L M, Kassim A and Gofar N 2011 Performances of two instrumented laboratory models for the study of rainfall in fi ltration into unsaturated soils *Eng. Geol.* 117 78–89

[11] Yunusa G H, Kassim A and Gofar N G 2014 Effect of soil layering on suction distribution in unsaturated residual soil slope *Electron. J. Geotech. Eng.* 19 9351–76

[12] Rahardjo H, Satyanaga A, Wang C L, Wong J L H and Lim V H 2018 Effects of unsaturated properties on stability of slope covered with Caesalpinia crista in Singapore *Environ. Geotech.*

[13] Rahardjo H, Satyanaga A and Leong E C 2013 Effects of flux boundary conditions on pore-water pressure distribution in slope *Eng. Geol.* 165 133–42
[14] Yunusa G H, Kassim A and Gofar N 2014 Effect of surface flux boundary conditions on transient suction distribution in homogeneous slope *Indian J. Sci. Technol.* **7** 2064–75
[15] Potts D M, Kovacevic N and Vaughan P R 1997 Delayed collapse of cut slopes in stiff clay *Geotechnique* 47 953–82
[16] Feng J, Gao L, Zhang L and Feng S 2019 Stability of railway embankment of China under extreme storms *Environ. Geotech.* 6 269–83
[17] Geo-Slope 2012 *Seepage Modeling with SEEP / W 2007* (Geo-Slope International, Ltd.)
[18] Geo-Slope 2020 *Stability modeling with GeoStudio* (Canada: Geo-Slope International, Ltd.)
[19] Fredlund D G and Rahardjo H 1993 *Soil mechanic for unsaturated soils* (Newyork: Wiley)
[20] ASTM-D2487-17e1 2017 *Standard Practice for Classification of Soils for Engineering Purposes (Unified Soil Classification System)* (West Conshohocken, PA)
[21] ASTM-D6836-02 2002 *Standard Test Methods for Determination of the Soil Water Chararcteristic Curve for Desorption Using a Hanging Column, Pressure Extractor, Chilled Mirror Hygrometer, and/or Centrifuge* (West Conshocken, PA)
[22] Rahardjo H, Kim Y and Satyanaga A 2019 Role of unsaturated soil mechanics in geotechnical engineering *Int. J. Geo-Engineering* 10 1–23
[23] Kristo C, Rahardjo H and Satyanaga A 2019 Effect of hysteresis on the stability of residual soil slope *Int. Soil Water Conserv. Res.* **7** 226–38
[24] Fredlund D G, Xing A, Fredlund M D and Barbour, S L 1995 The relationship of the unsaturated soil shear strength to the soil-water characteristic curve *Can. Geotech. J.* 32 440–8
[25] Rubin J and Steinhardt R 1963 Soil water relations during rain infiltration: I. theory *Soil Science Society Proceedings*
[26] Kassim A, Gofar N and Lee L M 2008 Numerical simulation of infiltration through unsaturated layered soil column *Malasian J. Civ. Eng.* 20 211–22
[27] Ng C W W, Wang B and Tung Y K 2001 Three-dimensional numerical investigations of groundwater responses in an unsaturated slope subjected to various rainfall patterns *Can. Geotech. J.* 1049–62
[28] Rahardjo H, Lee T T, Leong E C and Rezaur R B 2004 A flume for assessing flux boundary characteristics in rainfall-induced slope failure studies *Geotech. Test. J.* 27 145–53
[29] Gofar N and Lee L M 2008 Extreme rainfall characteristics for surface slope stability in the Malaysian Peninsular *Georisk Assess. Manag. Risk Eng. Syst. Geohazards* 2 65–78
[30] Gue S S and Tan Y C 2006 Landslides: case histories, lessons learned and mitigation measures *Conference on Landslide, Sinkhole, Structure Failure: MYTH or SCIENCE?* (Ipoh, Perak) pp 6–7
[31] Sutejo Y and Gofar N 2015 Effect of area development on the stability of cut slopes *Procedia Eng.* 125 331–7
[32] Sadeghi J M and Babaee A 2006 Structural optimization of B70 railway prestressed concrete sleepers *Iran. J. Sci. Technol. Trans. B, Eng.* 30 461–73
[33] Sadeghi J M 2015 *Ballasted railway tracks fundamentals of analysis and design* (Iran: Iran University of Schience and Technology)
[34] Gofar N and Lee L M 2009 PERISI : A computer program for preliminary evaluation of rainfall -induced slope instability *Malaysian J. Civ. Eng.* 21 235–40
[35] Yuan J, Papaioannou I and Straub D 2018 Probabilistic failure analysis of infinite slopes under random rainfall processes and spatially variable soil *Georisk Assess. Manag. Risk Eng. Syst. Geohazards* 1–14

[36] Seki, K. (2007) SWRC fit - a nonlinear fitting program with a water retention curve for soils having unimodal and bimodal pore structure. *Hydrol. Earth Syst. Sci. Discuss.*, 4: 407-437

IOP Conf. Series: Materials Science and Engineering 1153 (2021) 012007 doi:10.1088/1757-899X/1153/1/012007

[37] Bello Yamusa, Y., Azril Hezmi, M., Ahmad, K., Anuar Kassim, K., Sa'Ari, R., Alias, N., . . . Rashid, A. S. A. (2019). Soil water characteristic curves for laterite soil at different water contents and methods as lining system. *IOP Conference Series: Materials Science and Engineering*, , 527(1)

IOP Conf. Series: Materials Science and Engineering 1153 (2021) 012008 doi:10.1088/1757-899X/1153/1/012008

Effect of Biocementation via Enzymatic Induced Calcium Carbonate Precipitation (EICP) on the Shear Strength of Compacted Clay Liner

M U Zango[1,2], K A Kassim[1], A S Muhammed[1,3], K Ahmad[1], J Makinda[1,4]

[1]School of Civil Engineering, Universiti Teknologi Malaysia 81300 Johor Bahru, Johor, Malaysia
[2]Department of Civil Engineering, KUST, Wudil P.M.B. 3244 Kano State, Nigeria
[3]Department of Civil and Water Resources Engineering, Unimaid P.M.B. 1069 Maiduguri, Borno State, Nigeria
[4]Faculty of Engineering, Universiti Malaysia Sabah, Jalan UMS, 88400 Kota Kinabalu, Sabah, Malaysia
Email: muttaqaubaz@yahoo.com

Abstract. The strength of residual soil treated via biocementation means by employing enzymatic induced calcium carbonate precipitation (EICP) was assessed based on the standard recommended for compacted clay liner. EICP treated samples were prepared at four different concentrations of cementation solution (urea-$CaCl_2$) (0.25, 0.50, 0.75 and 1.00 M) and at various moulding water content (-2, 0, +2 and +4% OMC) using reduced British standard light compaction effort (RBSL). The result obtained has shown UCS values of untreated natural soil at the four molding water contents were less 200 kPa, the minimum standard recommended for compacted clay liner. Upon EICP treatment, it was determined that the UCS values increase with the increase in the concentration of cementation solution. The treated soils at all the cementation solutions and molding water contents have UCS values greater than 200 kPa. The lowest strength of the treated soils was 468.5 kPa determined at 0.25 M cementation solution and +2% OMC molding water content. The maximum UCS (643.5 kPa) value was determined at 1.00 M urea-$CaCl_2$ and -2% OMC. The results also reveal that calcium carbonate content in the treated soil increases with the increase in the concentration of cementation solution. Microstructural analysis on the treated soil indicated the presence of white precipitation within the pore space of the soil, and the mineral was confirmed to be calcite through XRD analysis.

1. Introduction

Compacted clay liner is one of the most essential components of engineered sanitary landfills provided to protect groundwater and native soil from the harmful effect of decomposed leachate [1]. Constructability in terms of shear strength is one of the four main criteria considered in the selection of suitable material for the construction of a liner [2]. Compacted clay liner should have adequate strength that is capable to sustain its own self-weight and expected solid waste. The shear strength of compacted clay liner should be able to withstand stresses that could develop from wet-dry or freeze-thaw cycles [3]. A value of 200 kPa unconfined compressive strength (UCS) had been recommended in literature by Daniel and Wu [4] as minimum strength suitable for compacted clay liner.

Not all natural soil meets the engineering requirements for the intended use. Thus, soil improvement is one of the most significant tasks of geotechnical engineering [5]. Considerable conventional materials (cement, lime and bentonite) and alternative materials (fly ash, rice husk ash, palm fuel ash) were investigated in the past to either substitute or amend natural native soil for the construction of compacted clay liner [1]. Although these materials were found to be effective in soil improvement. But, the majority of the investigated materials are threat to the environment due to either emission of greenhouse gases or release toxic by-product [6–8].

Accordingly, a new soil improvement technique known as bio-mediated soil improvement technique is being developed recently to reduce the negative environmental impacts posed by cement and other chemical stabilizers employed in soil improvement [9–11]. The technique uses either microorganisms or biological products to produce a chemical substance that would eventually improve the engineering properties of soil [10,12]. .

In EICP, free urease enzyme is used instead to trigger the formation of calcium carbonate precipitation via ureolysis [13]. The free urease in EICP can be extracted from an agricultural, microbial and fungal source. Agricultural urease is occurring in many high order plants such as sword beans, jack beans (Canavalia ensiformis), soybeans (Glycine max), melons, pine family, mulberry leaf, and pigweed [11].

Research into the use of EICP for civil engineering applications can be traced back to 2003 on the work conducted by Nemati and Voordouw [14]. They investigated the influence varying the strength of urease enzyme, urea and calcium chloride on the hydraulic conductivity of sandstone. Recent studies conducted on the improvement of shear strength of sandy soils using EICP process includes [15,16].

Hitherto, limited studies have been conducted on the application of MICP and EICP in fine-grained clay soils, specifically in improving engineering properties of compacted clay liner. Among the few works conducted on MICP with residual soils is that of [17,18]. This study is intended to improve the strength of residual clay soil for application in compacted clay liner at reduced British standard compactive effort.

2. Materials and Methods

2.1 Materials

The natural soil utilized in this work was sampled using a disturbed sampling method below a depth of about 1.5 m at Universiti Teknologi Malaysia (UTM) JB. The soil was reddish-brown and dominated by kaolinite clay minerals as revealed by XRD analysis.

The EICP solutions employed for this study are made up of free urease enzyme, urea and calcium chloride dihydrate. Four different equimolar concentration of urea-$CaCl_2$ (0.25, 0.5, 0.75 and 1.00 M) were used in the treatment process.

2.2 Methods

In order to determine the index properties of the natural soil, the laboratory tests that include particle size analysis, specific gravity and Atterberg limits (liquid limit, plastic limit, linear shrinkage, and plasticity index) were conducted based [19]. Reduced British standard light (RBSL) is selected in this study for the compaction test and subsequent unconfined compressive strength test. Compaction test using RBSL involves impacting 2.5 kg rammer falling from 300 mm on three layers of soil, each of which receives 15 number of blows. The test was performed on the natural residual soil sample that passed through a 20 mm size sieve as described in BS 1377 [20].

The EICP solution for the treatment was first made by mixing urea ($CO(NH_2)_2$) and calcium chloride dihydrate ($CaCl_2.2H_2O$) at various equimolar concentrations (0.25, 0.50, 0.75 and 1.00) with distilled and de-ionized water until all the solutes were dissolved. The free urease enzyme at 3 g/L was then added to the prepared urea-$CaCl_2$ solution and stirred for about 5 minutes to form EICP Solution. The EICP solutions were used to prepare UCS samples at various molding water contents relative to the optimum moisture content determined in the previous section. The mix and compact method was employed in preparing the UCS sample. The dried soils were mixed with EICP solutions and then compacted to various densities corresponding to the molding water contents -2%, 0%, +2% and +4% using static compaction machine. The extruded were then cured for 3 days in a humidity chamber that was operating at 99 RH and 25 °C.

The Unconfined Compressive Strength, UCS test was conducted by following the procedure described in [19]. The unconfined compressive test was performed by mounting the cured samples prepared in section 2.2.3 on the triaxial compression machine. The samples were compressed uniaxially until failure at a rate of 1.52 mm/min.

About 20 g of dried soil specimens, obtained from failed UCS samples were used for the determination of calcium carbonate content. The dried soils were soaked in a 4 M HCl acid for 24 hours. The mixture of the soil dissolved $CaCO_3$ and acid were washed thoroughly with distilled by using a 63 μm sieve. A filter paper was then used to separate the soil residue and the liquid . The soil residue was then dried in an oven and then weighed. The amount of calcium carbonate content in the treated soil was computed using the formula shown in equation 1, as adopted from Choi et. al [21].

$$CCC = 100 - \left(\frac{B}{A}\right) \times 100 \qquad (1)$$

CCC = Calcium carbonate content (%)
B = Mass of oven dried soil post washing
A = Mass of oven dried treated soil before washing

The scanning electron microscope (SEM) and X-ray diffraction (XRD) analyses were conducted on the natural and EICP treated soil in order to investigate the change in morphology and phases due to the formation of calcium carbonate precipitations and also to confirm the mineral content of the precipitation formed.

3. Results and Discussion

3.1 Index Properties of the Natural Soil

The physical and microstructural properties of the natural soil are presented in table 1. The soil is classified as sandy silts of high plasticity according to BSCS. The dominant clay mineral in the soil is kaolinite as revealed by XRD analysis on the natural soil. The soil was found to have a high liquid limit, plastic limit and plasticity index of 79, 30 and 49% respectively. The values of maximum dry density and optimum moisture content were 1.36 Mg/m^3 and 32% respectively. These values are later used in soil preparation for the unconfined compressive strength test.

Table 1. Physical Properties of the Soil

Property	Quantity
Natural Moisture Content (%)	32.72
Percentage Passing 63 μm Sieve (%)	57
Gravel Fraction (%)	24.16
Sand Fraction	17.16
Liquid Limit (%)	79
Plastic Limit (%)	30
Plasticity Index (%)	49
Linear Shrinkage (%)	16
Specific Gravity	2.63
BSCS Classification	Sandy silts soil of high plasticity
Maximum Dry Density (Mg/m^3)	1.36
Optimum Moisture Content (%)	32

3.2 Unconfined Compressive Strength (UCS)

The relationship between UCS and molding water contents for all the natural and treated soils are shown in figure 1. There is a general decrease in UCS with molding water in both natural soil and enzymatic-induced calcium carbonate (EICP) treated soils. At higher molding water contents, the soil structures are deflocculated due to the loss of cementation between soil particles, thus reduction in shear strength. A similar pattern of results was determined by Eberemu [1].

Figure 1 also revealed that the samples of the natural soil compacted at RBSL compaction energy for all the molding water content (-2% to +4% relative to OMC) have UCS less than 200 kPa, the minimum standard required for compacted clay liner. The compacted soils have attained the minimum value of shear strength (>200 kPa) upon treatment with EICP solution at 0.25 M to 1.00 M cementation solution. The increase in UCS due to EICP treatment is attributed to the formation of bio-cement in form of calcite carbonate precipitates between soil particles pore spaces [22,23]. As also depicted in figure 3, UCS increases with the increment in the concentration of cementation solution at any particular molding water content. Thus, the highest values of shear strength of 643.5, 573, 563.5, and 533 kPa are obtained at 1.00 M concentration of cementation solution when the samples were prepared at -2, 0, +2 and +4% moulding water content relative to OMC, respectively. The highest UCS values of 2.61±0.55 Mpa were also obtained at 1.0 M cementation solution by Deng and Wang [24] when sandy soil were treated by the MICP technique.

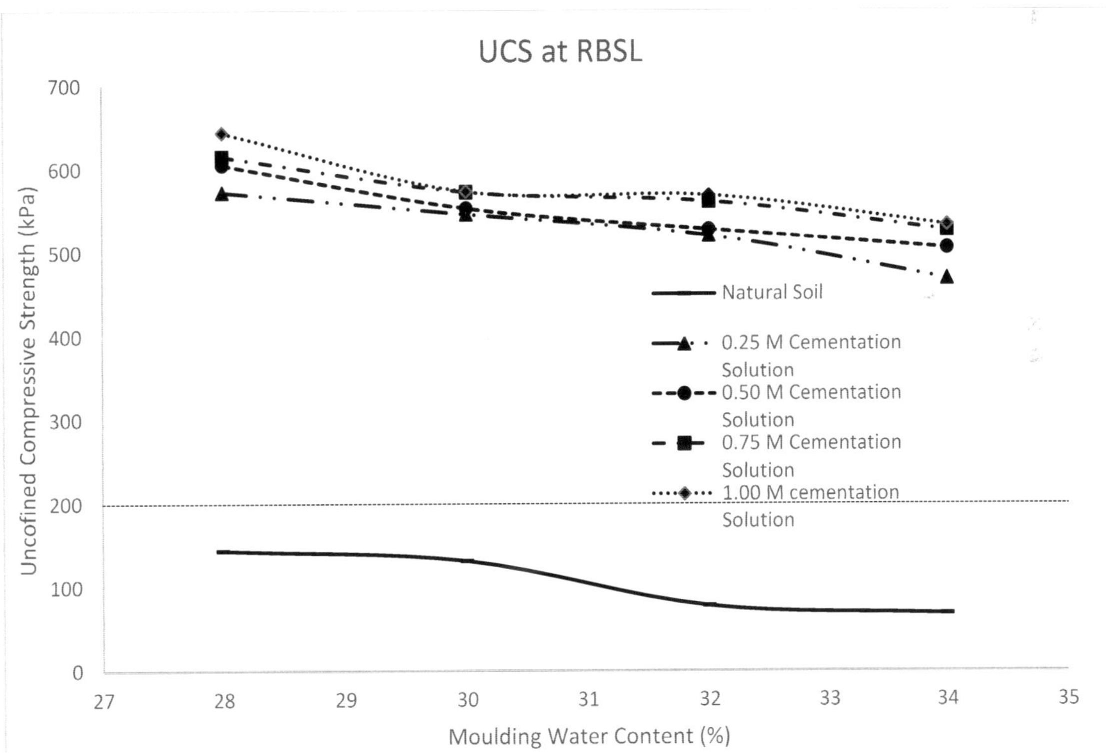

Figure 1. Variation of Unconfined Compressive Strength with Moulding Water Content in the Natural Soil and EICP Treated Soil

Figure 2 was plotted to explain the increase in calcium carbonate content in the EICP treated soil with the increment in cementation solution. As seen in the figure, calcium carbonate precipitations increase with a higher cementation solution from 0.25 to 1.0 M urea-$CaCl_2$. The figure also shows that; calcium carbonate content reduces with the increase in molding water content relative to the OMC. As explained by Okwadha and Li [25], the increase in calcium carbonate precipitation with the increment cementation solution is due to the increase in urea available for conversion to carbonate ions and also the presence of the high amount of calcium ions from calcium chloride. While the reduction in calcium carbonate content with the increment in molding water content relative to OMC is ascribed to the high amount of water that led to the less absorption of calcium carbonate in higher molding water content [26]. Osinubi et al. [27] reported a similar pattern of results.

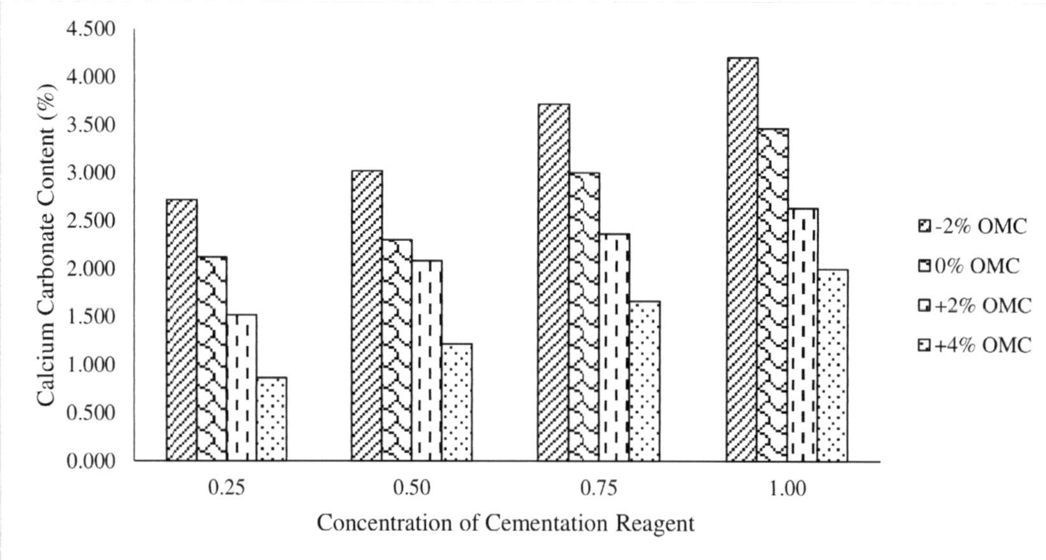

Figure 2. Effect Concentration of Cementation Solution on Calcium Carbonate Content in the Soil

3.3 Microstructure of the Treated Soil

The SEM and XRD analyses were performed on both natural and EICP treated soil in order to confirm the formation of bio-calcite within the treated soil phase. Figure 3 shows the SEM analysis of the natural and treated soil. Formations of white precipitations can be observed in the SEM image of the treated soil. XRD analysis conducted on the treated soil indicated the presence of calcite mineral as can be seen in figure 4.

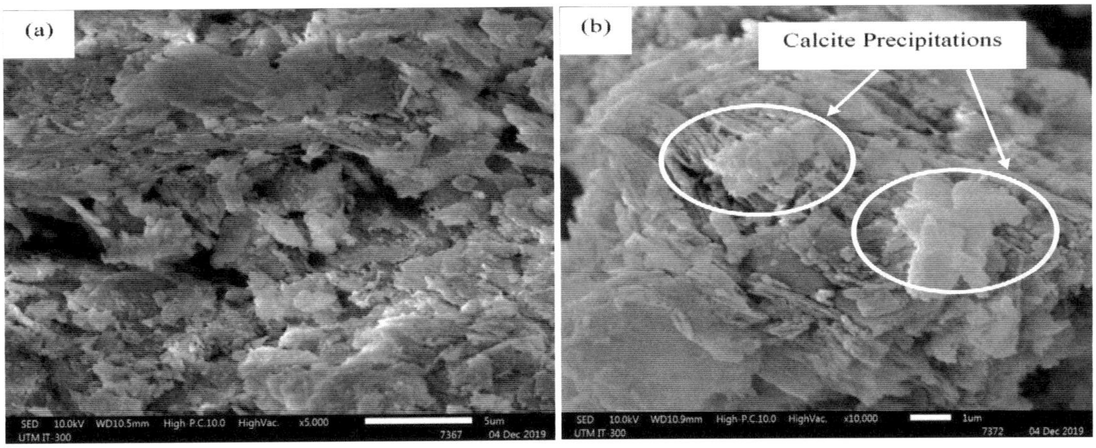

Figure 3. SEM Images of (a) Natural and (b) EICP Treated Soil

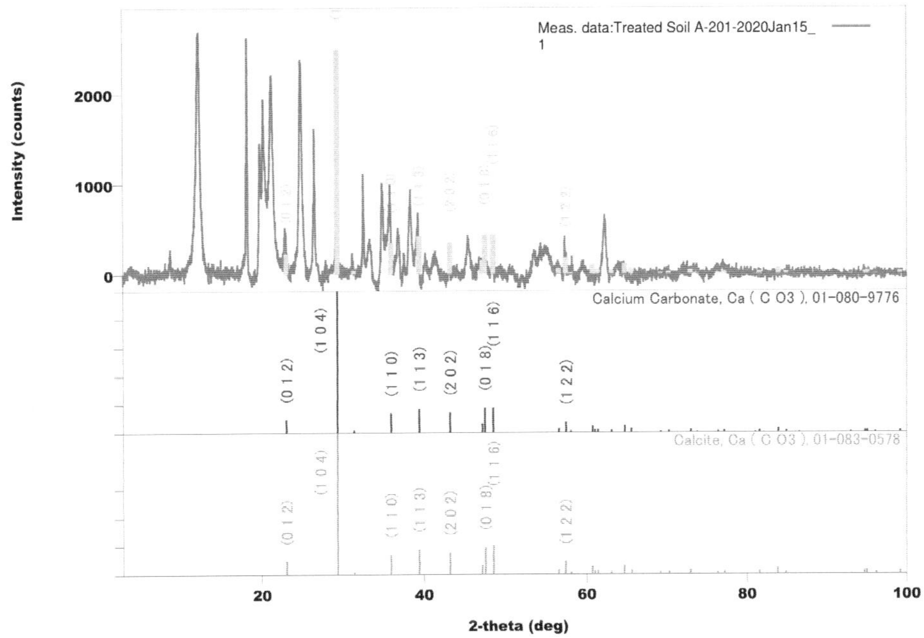

Figure 4. XRD of EICP Treated Soil

4. Conclusion and Recommendation

This study evaluated the strength of EICP treated residual clay soil relevant to compacted clay liner. The residual clay soil was treated with varying concentrations of cementation solution ranging from 0.25 to 1.00 M and the samples for unconfined compressive strength tests were prepared at 4 four different molding water contents relative to OMC (-2, 0, +2 and +4% OMC) using RBSL compaction effort. The results showed that the shear strengths of untreated residual soil 143.7 – 68.3 kPa were found to be less than 200 kPa recommended for CCL, but upon EICP treatment the shear strengths were found to be well above 200 kPa. The shear strengths were seen to increase with increment in the molarity of urea-$CaCl_2$ solution. The highest strength of 643.5 kPa was obtained at 1.00 M cementation solution when the sample was prepared at -2% moulding water content relative to OMC.

Finally, in order to recommend the utilization of EICP treated residual clay soil for the construction of liner or cover in a hydraulic barrier, other parameters such as hydraulic conductivity, volumetric shrinkage strength and leachate compatibility of the treated soil should be investigated.

Acknowledgement
The authors hereby recognize the financial support from grants Q.J130000.2522.20H21-GUP from UTM, R.J130000.7851.5F256-FRGS from Ministry of Higher education and Q.J130000.2451.04G57 from UTM. The first and third author also appreciate scholarship support from TETFUND-Nigeria.

References

[1] Eberemu A O 2013 Evaluation of bagasse ash treated lateritic soil as a potential barrier material in waste containment application *Acta Geotech.* **8** 407–21

[2] Benson B C H, Daniel D E and Boutwell G P 1999 Field Performance of Compacted Clay Liners *Geotech. Geoenvironmental Eng.* **125** 390–403

[3] Osinubi K J 2015 Influence of Molding Water Content on Shear Strength Characteristic of Compacted Cement Kiln Dust Treated Lateritic Soils for Liners and Covers *Niger. J. Technol.* **34**

[4] Daniel D E and Wu Y-K 1993 Compacted Clay Liners and Covers for Arid Sites *J. Geotech. Eng.* **119** 223–37

[5] Baldovino J A, Moreira E B, Teixeira W, Izzo R L S and Rose J L 2018 Effects of lime addition on geotechnical properties of sedimentary soil in Curitiba, Brazil *J. Rock Mech. Geotech. Eng.* **10** 188–94

[6] Rubinos D A and Spagnoli G 2018 Utilization of waste products as alternative landfill liner and cover materials – A critical review *Crit. Rev. Environ. Sci. Technol.* **0** 1–63

[7] Karol R H 2003 *Chemical Grouting and Soil Stabilization* vol 3

[8] Jason T. DeJonga, Mortensenb B M, Martinezb B C and Nelsonc D C 2010 Bio-mediated soil improvement *Ecol. Eng.* **36** 197–210

[9] Jawad Kadhim F and Zheng J-J 2016 Review of the Factors That Influence on the Microbial Induced Calcite Precipitation *Civ. Environ. Res.* **8** 69–76

[10] DeJong J T, Mortensen B M, Martinez B C and Nelson D C 2010 Bio-mediated soil improvement *Ecol. Eng.* **36** 197–210

[11] R.A.N. Dilrukshi S K 2016 Plant-derived urease induced sand cementation used in geotechnical engineering applications *International Conference on Geomechanics, Geo-energy and Geo-resources*

[12] Umar M, Kassim K A and Ping Chiet K T 2016 Biological process of soil improvement in civil engineering: A review *J. Rock Mech. Geotech. Eng.* **8** 767–74

[13] Putra H, Yasuhara H, Kinoshita N and Hirata A 2017 Optimization of Enzyme-Mediated

Calcite Precipitation as a Soil-Improvement Technique: The Effect of Aragonite and Gypsum on the Mechanical Properties of Treated Sand *Crystals* **7** 59

[14] Nemati M and Voordouw G 2003 Modification of porous media permeability, using calcium carbonate produced enzymatically in situ *Enzyme Microb. Technol.* **33** 635–42

[15] Putra H, Yasuhara H and Kinoshita N 2017 Optimum Condition for the Application of Enzyme-Mediated Calcite Precipitation Technique as Soil Improvement Technique *Int. J. Adv. Sci. Eng. Inf. Technol.* **7** 2145

[16] Almajed A, Tirkolaei H K, Kavazanjian E and Hamdan N 2019 enzyme Induced Biocementated sand with High strength at Low Carbonate Content *Sci. Rep.* **9** 1–7

[17] Umar M and Kassim K A 2017 Effect of Cementation Reagent on Biocementation of Tropical Residual Soil **90** 79–90

[18] Albuquerque D C, Cardoso R, Monteiro G A, Duarte S O, Martins V C and Cardoso S 2019 Towards a portable magnetoresistive biochip for urease-based biocementation monitoring *6th IEEE Port. Meet. Bioeng. ENBENG 2019 - Proc.* **4** 1–4

[19] BS 1377-2 1990 *British Standard Methods of test for Soils for civil engineering purposes. Part 2: Classification tests*

[20] BS 1377-4 1990 *British Standard Methods of test for Soils for civil engineering purposes. Part 4: Compaction-related tests* (United Kingdom)

[21] Choi S-G, Park S-S, Wu S and Chu J 2011 Methods for Calcium Carbonate Content Measurement of Biocemented Soils *J. Mater. Civ. Eng.* **29** 06017015

[22] Cheng L, Shahin M A, Asce M and Mujah D 2013 Influence of Key Environmental Conditions on Microbially Induced Cementation for Soil Stabilization 1–11

[23] Li M, Fang C, Kawasaki S and Achal V 2018 Fly ash incorporated with biocement to improve strength of expansive soil *Sci. Rep.* **8** 4–10

[24] Deng W and Wang Y 2018 Investigating the Factors Affecting the Properties of Coral Sand Treated with Microbially Induced Calcite Precipitation *Adv. Civ. Eng.* **2018** 1–7

[25] Okwadha G D O and Li J 2010 Optimum conditions for microbial carbonate precipitation *Chemosphere* **81** 1143–8

[26] Choi S G, Chu J, Brown R C, Wang K and Wen Z 2017 Sustainable Biocement Production via Microbially Induced Calcium Carbonate Precipitation: Use of Limestone and Acetic Acid Derived from Pyrolysis of Lignocellulosic Biomass *ACS Sustain. Chem. Eng.* **5** 5183–90

[27] Osinubi K J, Gadzama E W, Eberemu A O, Ijimdiya T S and Yakubu S E 2019 Evaluation of the Strength of Compacted Lateritic Soil Treated with Sporosarcina Pasteurii *ICEG 2018: Proceedings of the 8th International Congress on Environmental Geotechnics* vol 3, ed L Zhan, Y Chen and A Bouazza (Springer Singapore) pp 419–28

IOP Conf. Series: Materials Science and Engineering 1153 (2021) 012009 doi:10.1088/1757-899X/1153/1/012009

Modelling the applicability of Low Impact Development (LID) technologies in a university campus in the Philippines using Storm Water Management Model (SWMM)

R A Frias and M Maniquiz-Redillas
Hydraulics and Water Resources Engineering, Department of Civil Engineering, John Gokongwei Jr., College of Engineering, De La Salle University, 2401 Taft Avenue, Manila Philippines
E-mail: marla.redillas@dlsu.edu.ph

Abstract. Progressive land development increases the imperviousness of an area which disrupts the water balance and results to the degradation of water quality, high peak flow, and excessive volume of surface runoff. One efficient approach to address this issue on water is the application of Low impact development (LID) technologies. LID helps improve the water quality and water quantity of an area to maximize its land-use. In this study, the eight LID technologies available in SWMM namely, bio retention cell, infiltration trench, rain garden, green roof, permeable pavement, rain barrel, rooftop disconnection, and vegetative swale were applied in simulating the area of the De La Salle University - Laguna Campus, a pre-developed area in Binan, Laguna which currently undergoes land-use change. This study area was simulated without the use of LID, with the use of varying LID, and capturing the 80, 90, and 95 percent of all the rainfall from 1989-2018. Digital elevation model and disaggregated rainfall data were used as input in SWMM. The ArcMap 10.4 was utilized to delineate and produce the three sub catchments with an area of 24.13 ha., 10.18 ha., and 4.34 ha., respectively while Storm Water Management Model (SWMM) software was used in analyzing these sub catchments to produce water balance values. Results showed that more than 60 percent of the disaggregated rainfall was under the one-year return period. Also, bioretention cell with 80th percentile rainfall maximizes the reduction of runoff while infiltration trenches with 95th percentile rainfall were the most effective in increasing the infiltration among the eight LID technologies. It was also indicated that the surface runoff in first sub catchment was reduced to more than 80% using bioretention cell, infiltration trench, rain garden, or rooftop disconnection. Finally, the area of a sub catchment has a positive correlation in its reduced runoff while a negative correlation in its infiltration when integrating the LID technologies. This research can be a resource for further studies and in to support SDGs 6, 9, and 11 to have a better water management, resilient infrastructures, and attain sustainable cities and communities.

1. Introduction

Simultaneous with the rapid increase of population and urbanization is land development. It is an ingredient and indicator for the progress of a developing city or country [1]. These include the erection of more infrastructures such as roads, bridges, highways, buildings, and power supplies which creates more impervious areas and reduces green spaces. Imperviousness is correlated to a faster time of concentration, a higher volume of peak flow rates, and an increase of pollutants to water bodies [2]. Moreover, it contributes to the impacts of climate change and the destruction of the natural ecosystem [3].

In the Philippines, almost half of the Filipino lived in urban and cities, [4] and its capital, Manila is considered the world's most densely populated city. Urban sprawl is evidently seen which increases the impacts of rapid urbanization and climate change. These impacts include water scarcity and shortages, urban flooding, and degradation of water quality [5] which were evidently seen in the aftermath of the recent super typhoons that struck the Philippine archipelago. Additionally, Greenpeace (2007) stated nine water-critical urbanized area which includes Metro Manila, Metro Cebu, Davao, Baguio City, Angeles City, Bacolod City, Iloilo City, Cagayan de Oro City, and Zamboanga City. These areas will suffer from a lack of supply of fresh water due to increasing demand every year and the worsening of EL Niño or drought.

Currently, programs of the government focus on improving the existing drainage systems. However, these drainage systems only address the water quantity, not water quality and the implementation of sustainable development ideas and concepts in metropolitan urban areas is important to mitigate the impacts of these water issues and to the environment [6]. One approach is the Low impact development (LID) [7]. LID is a stormwater management approach that addresses the rainfall at the source and uses a micro-scale and mimics a site predevelopment hydrology by using design techniques that infiltrate, filter, store, evaporate and detain runoff controls [8]. Some types of LID are bioswales, rain gardens, green roofs, permeable pavements, and bioretention cells [9]. Bioswales are effective in capturing runoff from unusual rainfall event [10], proper biological design of rain gardens helps in reducing non-point source (NPS) pollution and maintains its efficiency through time [11], green roofs efficiently decrease the urban heat island effect on an area [12], fully permeable pavements minimizes the negative effects of stormwater runoff [13] while bioretention cells help in infiltrating stormwater from roads and parking lots and contributes in metal removal [14]. These LID technologies were also used in developed countries such as United States, South Korea, China, Australia, and UK namely, best management practices (BMPs), smart city, sponge cities, water-sensitive urban design (WSUDS), and sustainable urban drainage systems, respectively [15] and economically cost less than conventional stormwater management [16][17].

Geographic information system (GIS) is a framework for gathering, managing, and analyzing spatial data and organizes it into layers of information and was used to subdivide the sub catchments [18]. It helps in visualizing maps and 3D scenes using patterns, relationships, and situations of an area [19]. This is also utilized in different hydraulic engines, one of which is the Stormwater Management Model (SWMM). SWMM is used throughout the world in processing hydrologic phenomena in urban and non-urban areas. Different LID technologies which include rain gardens, bio-retention cells (bioswales), vegetative swales, infiltration trenches, green roofs, rooftop (downspout) disconnection, rain barrels or cisterns (rainwater harvesting), and continuous permeable pavement systems were included in the software. It is the only existing free software that simulates both water quantity and water quality [17]. Moreover, studies in South Korea used SWMM in assessing the hydrologic impact [20] and a tool for assessing flood impact [21]. Geomorphologic characteristics of watersheds as parameters in analyzing pre-developed conditions were used and a pre-developed state is more effective for the construction of LID technologies in terms of efficiency and cost [22].

In this study, the main objective is to determine the applicability and effectiveness of LID technologies in DLSU Laguna Campus using SWMM by comparing the performance of a conventional drainage system and a drainage system with LID. An 80th, 90th, and 95th percentile of rainfall was considered with the varying LID technologies in producing the water balance.

The findings of this study could be useful for further investigation and application of LID technologies in the campus to address the excess surface runoff and maximize the use of limited space.

2. Methodology

2.1. Site Description

DLSU Laguna Campus is a 52-ha campus located in Biñan, Laguna (14° 15' 46.64" N, 121° 02' 34.30" E, Figure 1) which has an annual rainfall of 944 mm. The dry season is from December 27 to May 21 while the wet season is from May 21 to December 27. The topography within 3.21 km of Biñan is essentially flat and covered by artificial surfaces (58%) and water (31%) [23]. At present, the DLSU

Laguna Campus occupies mostly green spaces or areas with pervious surfaces and envisioned to be the center of research and technology in the Philippines by which construction of more infrastructures are necessary. It is also located on the southwestern part of Laguna de Bay, the largest lake in the Philippines. Improvements on the drainage network are in the early phase and the application of LID technologies will minimize the effects of future urban flooding, high volume, and peak flow rates in its area and surrounding water bodies [24].

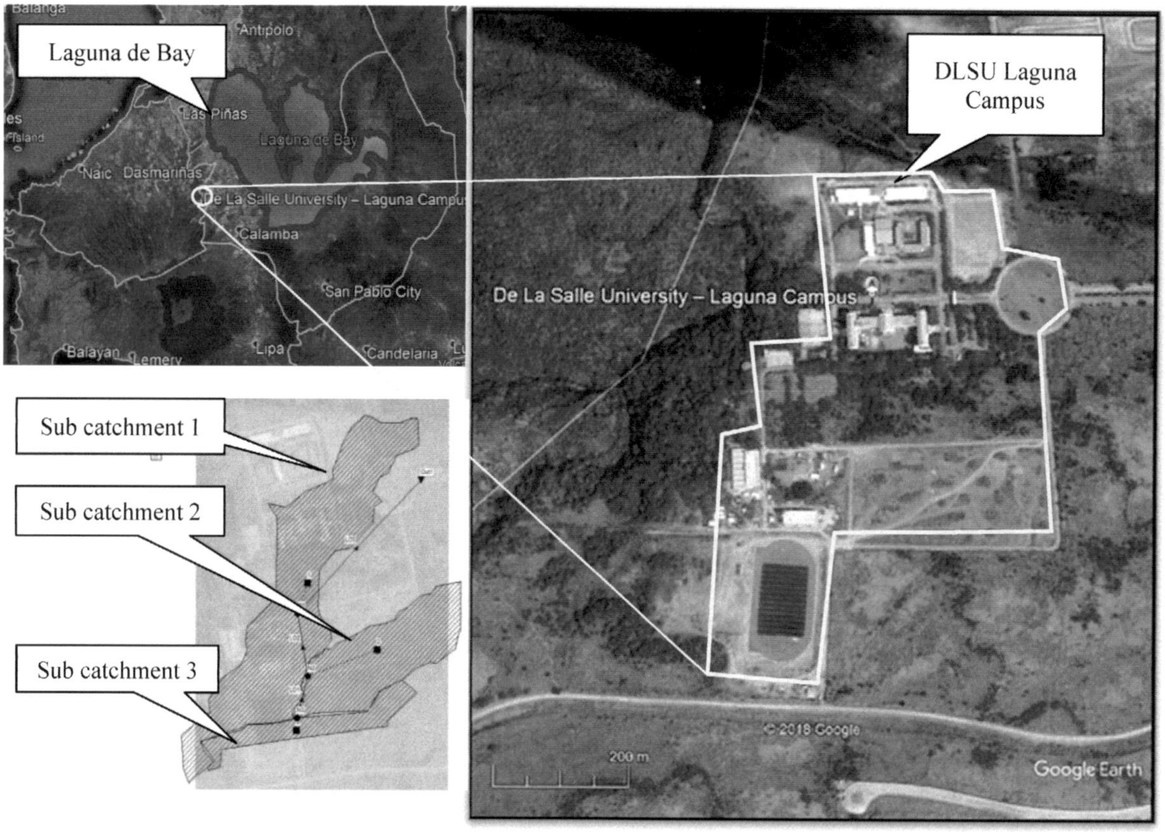

Figure 1. The DLSU Laguna Campus (right) with its delineated sub catchments (lower left) modelled from ArcMap 10.4 and SWMM 5.1 and Laguna de Bay (upper left).

2.2. Research Method

Shown in Figure 2 is the flowchart of research methodology which comprises of five parts namely, data gathering, processing of data, modeling and simulations, results of SWMM, and analysis of results.

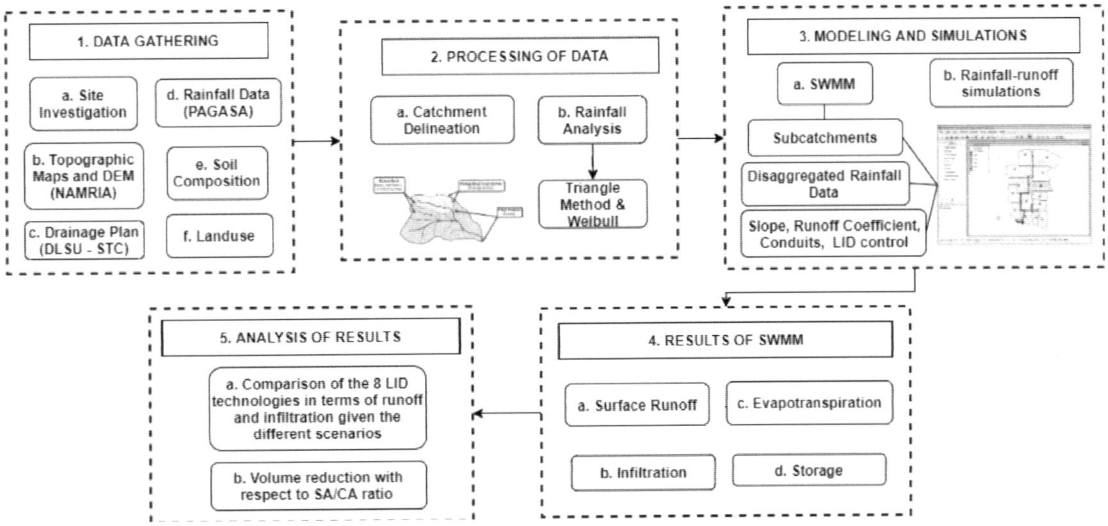

Figure 2. Flow chart of the research methodology.

2.2.1. Data Gathering.
The data gathered for this research include a 30-year daily rainfall data from 1989-2018 of Ambulong, Batangas and a 28-year rainfall data from 1991-2018 of NAS-UPLB, Laguna which were the nearest available rain gage stations to DLSU Laguna Campus in Biñan, Laguna. These data were acquired from Philippine Atmospheric, Geophysical and Astronomical Services and Administration (PAGASA) and all data mark with blank, '- ', or '-2' value are considered no data. Also, one-day rainfall data was considered one rainfall event though multiple rainfall events may occur in one day.

For the Digital Terrain Model acquired from the National Mapping and Resource Information Authority (NAMRIA) in the form of raster files in .tiff format, a 10 km radius was extended around the perimeter of the actual site. Other parameters such as slope, runoff coefficient, and soil composition were based on the topographic and land use map.

2.2.2. Processing of Data.
All raw data from the different government agencies were processed. The DEM file of DLSU Laguna Campus from NAMRIA was delineated using ArcMap 10.4 to produce sub catchments [25]. The rainfall data from PAGASA of the given two rain gage stations were calculated using Inverse Distance Weighting (IDW) method, equation (1) to produce a combined single rainfall data. It was disaggregated using the triangle method to produce hourly data from daily rainfall data [26] and plotted using Weibull plotting position formula, equation (2) to calculate the 80th, 90th, and 95th percentile [27].

$$Z_j = \frac{\Sigma_i\left(\frac{z_i}{d_{ij}^n}\right)}{\Sigma_i\left(\frac{1}{d_{ij}^n}\right)} \qquad (1)$$

Where,
Zi = the value of known point
dij = the distance to known point
Zj = the unknown point
n= a user selected exponent (often 1, 2 or 3)

$$P = \frac{m}{N+1} \tag{2}$$

Where,
P = probability of occurrence
N = total number of years recorded
m = the rank of descending values with the largest equal to 1

2.2.3. Modelling, Simulation, and Analysis of Results.
EPA SWMM 5.1 was used in the simulation [25]. All processed data including soil composition, runoff coefficient, curve number, and slope were used as input. After applying all the parameters of the SWMM model shown in Table 1, the disaggregated rainfall data for the 80th, 90th and 95th percentile was included. The model has been set and the simulation was conducted. Simulation without the use of LIDs served as a point of comparison. The variables are the varying percentiles and the eight different LIDs available in SWMM 5.1 for the three sub catchments.

Table 1. Characteristics of the sub catchments and parameters used in SWMM.

Sub catchment	Area (ha)	% Imperviousness	Width (m)	Curve Number, CN	Node, N	Invert elevation (m)	Link	Length (m)	Manning's Roughness
1	21.13	20	182	80	1	90	LO1	107	0.013*
2	10.18	10	258	80	2	94	LO2	140	0.013*
3	4.34	25	151	80	3	98	LO3	463	0.013*

Note: *Manning's roughness for concrete pipes.

Each sub catchment has a scenario by which the eight different LID technologies and a simulation without LID were applied and all these scenarios were simulated to the three different percentiles. Overall, a total of 81 scenarios were conducted for the three sub catchments.

Since evapotranspiration is negligible, the results focused on the total infiltration, and total surface runoff. A comparison between an analysis without LID and an analysis with each of the 8 LID technologies were calculated. Also, the volume reduction with respect to surface area/catchment area (SA/CA) ratio was presented. It was produced based on the designed total surface area of each LID technology on the area of each sub catchments.

3. Results and Discussion

3.1. Rainfall Characteristics
Distant rain gauges stations were common in developing countries like the Philippines. Rainfall data need to be analyzed first using statistical models to have a compatible input to different softwares like ArcMap 10.4 and SWMM 5.1[25]. Figure 3 shows the average monthly rainfall and cumulative frequency of the processed 30-year rainfall data from Ambulong, Batangas and 28-year rainfall from NAS-UPLB Laguna rain gage stations with their standard deviations included. February has the lightest monthly rainfall with an average total accumulation of 28.47 mm while the heaviest rainfall occurs in July (344.61 mm) with the largest standard deviation of 186.59 mm. The total accumulated annual rainfall is 1855.32 mm.

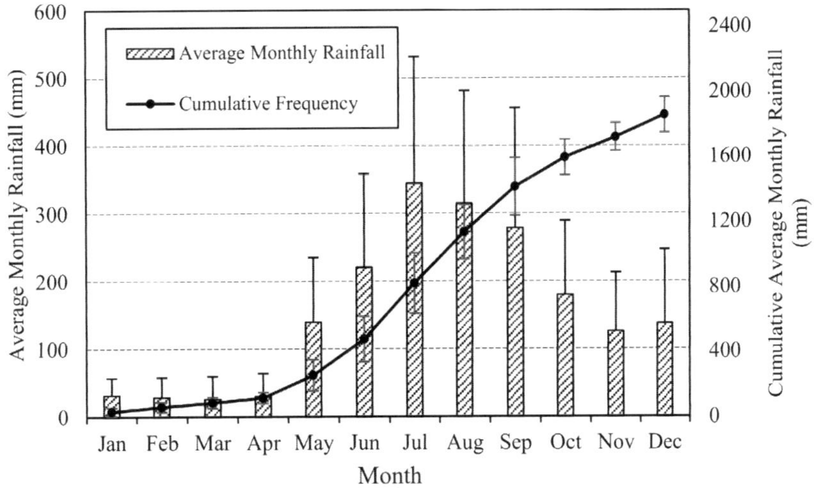

Figure 3. Average and cumulative monthly rainfall data with their standard deviation from the processed rainfall data from PAGASA (1989-2018).

The 80th, 90th, and 95th percentile were used in this study. As observed in Figure 4 (a), a visible change in slope was seen at around 0.8 and considered the minimum significant value. The value at 0.9 was used as the middle value and the value at 0.95 was used as the maximum significant value. The value near 1 to 1 was not considered as the maximum value because changes in that area were very minimal. Also, the 95th rainfall percentile event is based on the Section 438 Technical Guidance of USEPA [28]. Figure 4 (b) depicts the trendline of the disaggregated rainfall data with respect to the return period. The cost of the urban drainage design could be minimized by using the optimal rainfall return period [29].

The trendline was used in Figure 4 (c) which showed that more than 60% of rainfall were found in the 1-yr return period, less than 5% of the total rainfall were found in the 2 and 5-yr return period, and more than 30% of the total rainfall were found in the 10 and 20-yr return period.

An hourly time series with respect to rainfall depth for each respective 80th, 90th, and 95th percentile occurrence frequency disaggregated from daily rainfall data covering a period from 1989 to 2018 was shown in Figure 5. The values of these 80th, 90th, and 95th percentile were 4.80 mm, 15.10 mm, and 28.89 mm, respectively. This method was used as an alternative due to the lack of hourly rainfall data in the study area. The peak flow is always around the middle of the day and proportionately distributed throughout the day. This was used as input in the SWMM simulation.

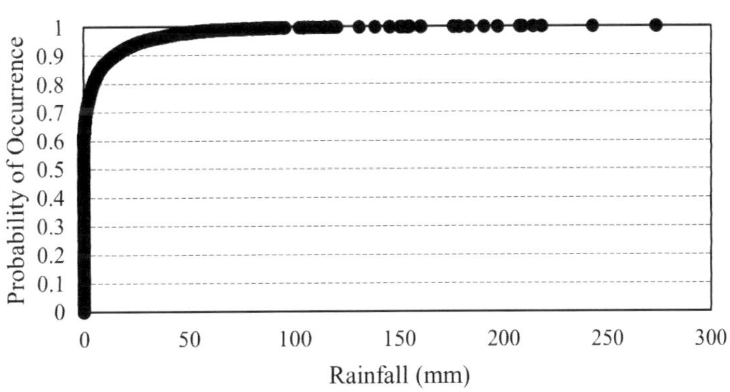

(a)

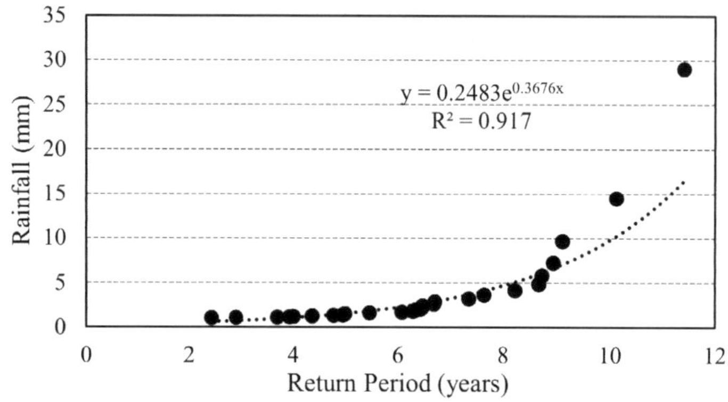

(b)

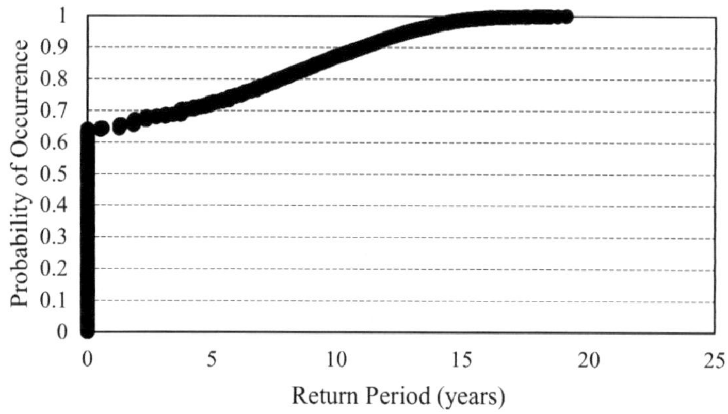

(c)

Figure 4. (a) Cumulative distribution function (CDF) of the disaggregated rainfall data from Ambulong, Batangas and NAS – UPLB Laguna rain gage stations, (b) Trendline of the disaggregated rainfall data, and (c) CDF of the disaggregated rainfall data with respect to return period using Weibull plotting position.

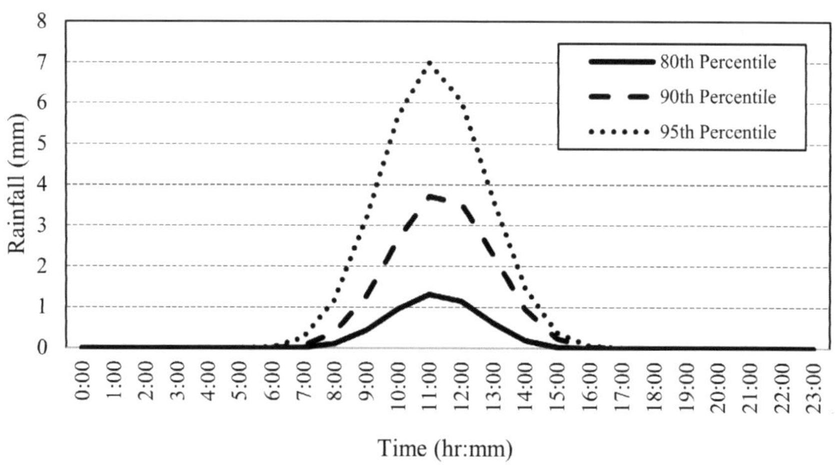

Figure 5. Hourly time series of rainfall depth of the 80th, 90th, and 95th percentile occurrence frequency of the disaggregated rainfall data (1989 – 2018) in DLSU Laguna Campus

3.2. *Water balance of the different scenarios of each sub catchments*

Shown in Figure 6 (a) were the water balances of the different scenarios of the simulation without the use of LID and using the eight LID technologies installed in SWMM. The 80th, 90th, and 95th percentile were used on each of the scenarios. There is a total of 27 scenarios in sub catchment 1 which has an area of 21.13 hectares by which NL stands for No LID, BC for bioretention cells, IT for infiltration trenches, RG for rain garden, GR for green roof, PP for permeable pavement, RB for rain barrel, RD for Rooftop disconnection, and VS for vegetative swale. Results showed that BC, IT, and PP created the most significant decrease in total runoff [30] while IT and PP significantly increase the infiltration of the area in the different percentiles. Poor performance was found in GR and VS in reducing the runoff and BC, RG, GR, RB, RD, and VS in increasing the infiltration in the three percentiles. Among the three percentiles, the 80th percentile has the most reduction in total runoff and the 95th percentile for the highest increase in the percentage of infiltration.

Figure 6 (b) shows the water balances of different scenarios same as in Figure 6 (a) but applied in sub catchment 2 with an area of 10.18 hectares. Also, there were a total of 27 scenarios and the same legend was applied. Results showed that LID technologies, including BC, IT, PP, and RD have the most significant total runoff reduction [31] and varying results for the increase of infiltration. For the 80th percentile, GR, and RB for increasing the infiltration and for the 90th and 95th percentile, PP showed a better result for infiltration. Among the three percentiles, same as in sub catchment 1, the 80th percentile has the most reduction in total runoff and the 95th percentile for the highest increase in the percentage of infiltration.

In Figure 6 (c), the water balances of the different scenarios applied in sub catchment 3 which has an area of 4.34 ha. were shown. Same scenarios and legend were used as in sub catchment 1 and 2. Also, there is a total of 27 scenarios. Results showed that BC, IT, and RG were effective in runoff reduction, and IT in increasing the infiltration on the different percentiles. Same trend as in the sub catchment 1 and 2 for the three percentiles, in the 80th percentile showed the most significant decrease in total runoff and in the 95th percentile showed the most significant increase in infiltration.

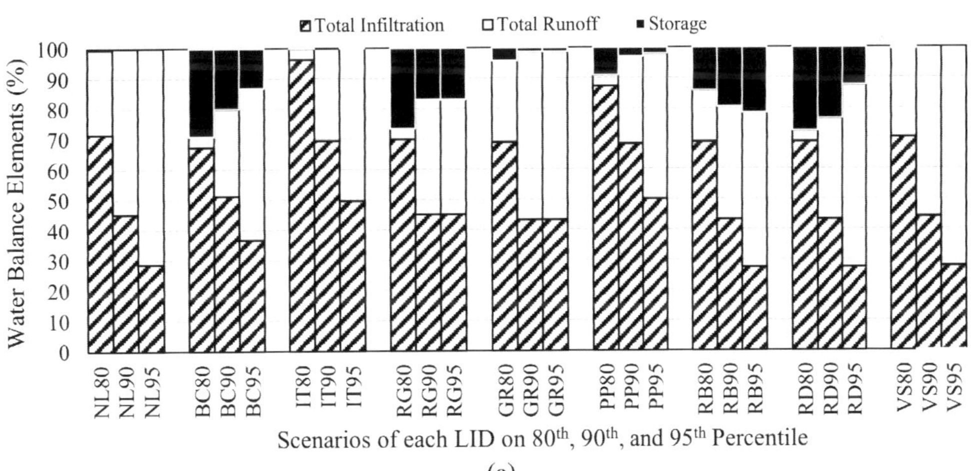

(a)

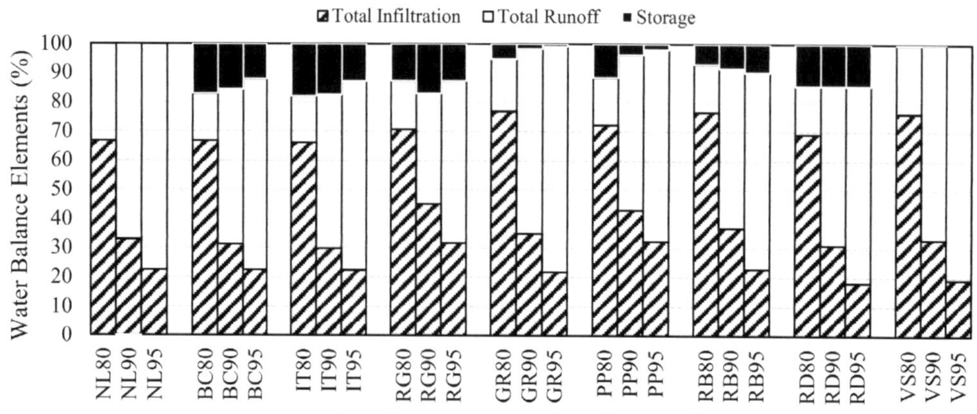

(b)

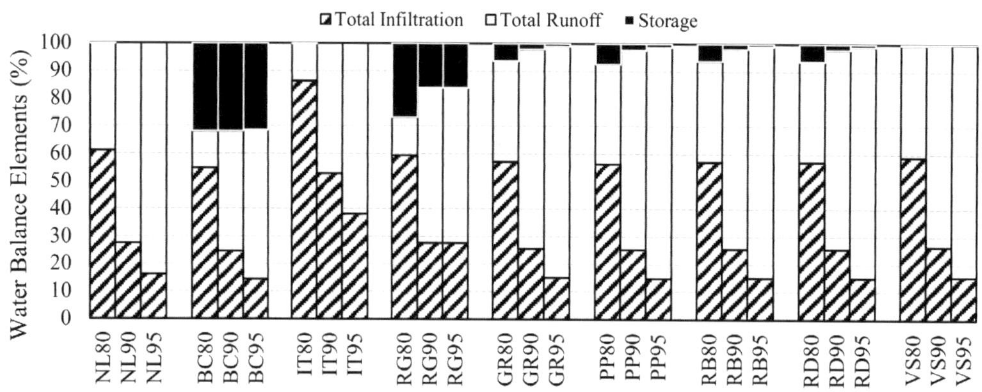

(c)

Figure 6. Comparison of the water balances of the scenarios with no LID (NL), with bioretention cell (BD), with infiltration trench (IT), with rain garden (RG), with green roof (GR), with permeable pavement (PP), with rain barrel (RB), with rooftop disconnection (RD), and with vegetative swale (VS) on the 80th, 90th, and 95th percentile rainfall of (a) sub catchment 1, (b) sub catchment 2, (c) sub catchment 3.

3.3. Efficiency of LID

The SA/CA ratio was an important factor in evaluating the performance of an LID in the reduction of runoff and pollutant loads [32]. Shown in Figure 7 is the reduced runoff or the equivalent increase in infiltration and storage in the 3 sub catchments vs SA/CA ratio [33]. It was produced from the SWMM model on each scenario was all less than 10% but effectively reduce the runoff on selected LID technology. The highest reduced runoff was found on the 80th percentile of sub catchment 1. These are the BC, IT, RG, and RD which reduced the runoff to 87%. The values found near 0 to 0 were results from the 90th and 95th percentile of sub catchment 3. These are the GR, PP, RB, RD, and VS. Also, results showed that reduced runoff in sub catchment 1 on the different LIDs, which has the largest surface area among the 3 sub catchments, have the highest reduced runoff.

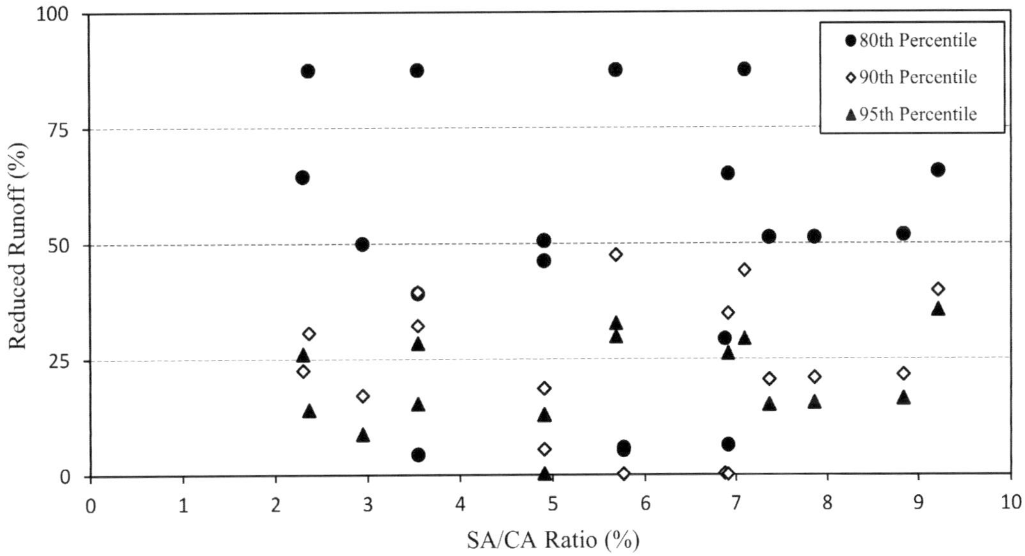

Figure 7. Reduced runoff of the 80th, 90th, 95th percentile rainfall with respect to Surface Area/Catchment Area (SA/CA) ratio.

4. Conclusion

Being sustainable is an essential ingredient in the progress of developing countries like the Philippines. Part of it is improving the water system of an area which includes controlling the excess water in the form of surface runoff, utilizing it for alternative use, and improving its water quality to minimize the degradation of water quality. One innovative approach is the use of LID technologies but challenges like insufficiency in data lead us to adapt other statistical methods to model and simulate different scenarios in an area unlike the methods used by developed countries. In this study in DLSU Laguna Campus located in Binan, Laguna, a developing city, the researcher focused on water quantity. Results showed that more than 60% of the rainfall data were within the one-year return period. Also, there is a significant increase in infiltration in using infiltration trenches and an effective decrease in surface runoff in using bioretention cells compared to the other available LID technologies in SWMM. In this site, using the 80th percentile rainfall was also best in maximizing the effectivity of LID technologies compared to the 90th and 95th percentile rainfall. The sub catchments with at least 10 hectares had a greater amount of reduced runoff and a lesser amount of infiltration. Also, applying the LID technologies does not require a large surface area and would mean a lesser cost. It is recommended to have an empirical data to support and improve the simulated results. This study serves as a pioneering research work and reference for future researchers interested in exploring LID applications not only in the Philippines but also in other developing countries.

Acknowledgement

This research was financially supported by the Department of Science and Technology – Engineering Research and Development for Technology (DOST – ERDT) scholarship program. The authors were grateful for their support.

References

[1] Orenstein D E and Hamburg S P 2010 Population and pavement: population growth and land development in Israel *Population and Environment* **31(4)** 223–254

[2] Minnesota Pollution Control Agency 2018 *Design Infiltration Rates* Retrieved from Minnesota Stormwater Manual: https://stormwater.pca.state.mn.us/index.php?title=Design_infiltration_ rates

[3] Esra A 2013 Impacts of Improper Land Uses in Cities on the Natural Environment and Ecological Landscape Planning Advances in Landscape Architecture doi:10.5772/55755

[4] Plecher H 2020 *Urbanization in the Philippines 2019* Retrieved https://www.statista.com/statistics/455910/urbanization-in-philippines/
[5] Indo-Asian News Service (IANS) 2017 *The News Minute* Retrieved from Urban flooding caused by plastic clogging, poor drainage: Experts.
[6] Surya B, Ahmad D N A, Sakti H and Sahban H 2020 Land Use Change, Spatial Interaction, and Sustainable Development in the Metropolitan Urban Areas, South Sulawesi Province, Indonesia *Land* **9(3)** 95 doi:10.3390/land9030095
[7] Hager J, Hu G, Hewage K, and Sadiq R 2018 Performance of low-impact development best management practices: a critical review *Environmental Reviews* **27(7)** Retrieved from https://www.nrcresearchpress.com/doi/abs/10.1139/er-2018-0048#.XTawaugzZPY
[8] United States Environmental Protection Agency (US EPA) 2018 Urban Runoff: Low Impact Development Retrieved from Polluted Runoff: Nonpoint Source (NPS) Pollution: https://www.epa.gov/nps/urban-runoff-low-impact-development
[9] Kaykhosravi S, Khan U T and Jadidi A 2018 A Comprehensive Review of Low Impact Development Models for Research, Conceptual, Preliminary and Detailed Design Applications *Water* **10(11) 1541** doi:10.3390/w10111541
[10] Monrabal-Martinez C, Aberle J, Muthanna T, and Orts-Zamorano M 2018 Hydrological benefits of filtering swales for metal removal *Water Research* **145** 509–517. doi: 10.1016/j.watres.2018.08.051
[11] Hong J, Geronimo F K, Choi H and Kim L-H 2018 Impacts of nonpoint source pollutants on microbial community in rain gardens *Chemosphere* **209** 20-27 doi: 10.1016/j.chemosphere.2018.06.062
[12] Yang J and Bou-Zeid E 2019 Scale dependence of the benefits and efficiency of green and cool roofs. *Landscape and Urban Planning.* **185**, 127–140. doi: 10.1016/j.landurbplan.2019.02.004
[13] Saadeh S, Ralla A, Al-Zubi Y, Wu R and Harvey J 2019 Application of fully permeable pavements as a sustainable approach for mitigation of stormwater runoff *International Journal of Transportation Science and Technology* **8(4)** 338-350 doi:10.1016/j.ijtst.2019.02.001
[14] Costello D M, Hartung E W, Stoll J T and Jefferson A J 2019 Bioretention cell age and construction style influence stormwater pollutant dynamics *Science of the Total Environment* **712** 135597 doi: 10.1016/j.scitotenv.2019.135597
[15] Fletcher T D, Shuster W, Hunt W F, Ashley R, Butler D, Arthur S, … Viklander M 2014 SUDS, LID, BMPs, WSUD and more – The evolution and application of terminology surrounding urban drainage *Urban Water Journal* **12(7)** 525–542. doi:10.1080/1573062x.2014.916314
[16] Penniman D, Hostetler M, Borisova T and Acomb G 2013 Capital Cost Comparisons between Low Impact Development (LID) and Conventional Stormwater Management Systems in Florida *Suburban Sustainability* **1(2)** http://dx.doi.org/10.5038/2164-0866.1.2.1
[17] US EPA 2017 Stormwater Management Model (SWMM) Retrieved from Environmental Protection Agency Website: https://www.epa.gov/water-research/storm-water-management-model-swmm
[18] Taji S G and Regulwar D G 2019 LID coupled design of drainage model using GIS and SWMM *ISH Journal of Hydraulic Engineering*, 1–14. doi:10.1080/09715010.2019.1660919
[19] Rai P K, Chahar B R and Dhanya C T 2016 GIS-based SWMM model for simulating the catchment response to flood events *Hydrology Research* **48(2)** *384–394.* doi:10.2166/nh.2016.260
[20] Jang S, Cho M, Yoon J, Yoon Y, Kim S, Kim G and Aksoy H 2007 Using SWMM as a tool for hydrologic impact assessment *Desalination* **212(1–3)** 344-356
[21] Ly S, Shin H, Kang N and Kim, H S 2014 Using SWMM as a Tool for Flood Impact Assessment *Korean Society of Hazard Mitigation* **(1)** 145 Retrieved from http://search.ebscohost.com/login.aspx?direct=true&db=edskyo&AN=edskyo.4010024144327&site=eds-live

[22] Kong F, Ban Y, Yin H, James P and Iryna D 2017 Modeling stormwater management at the city district level in response to changes in land use and low impact development *Environmental Modelling & Software* **95** 132-142

[23] Weather spark Average weather in Binan, Philippines 2021 Retrieved from https://weatherspark.com/y/136844/Average-Weather-in-Bi%C3%B1an-Philippines-Year-Round

[24] Jia H, Lu Y, Yu S L, and Chen Y 2012 Planning of LID–BMPs for urban runoff control: The case of Beijing Olympic Village *Separation and Purification Technology, 84, 112–119.* doi: 10.1016/j.seppur.2011.04.026

[25] Rai P K, Chahar, B R and Dhanya C T 2016 GIS-based SWMM model for simulating the catchment response to flood events *Hydrology Research* **48(2)** *384–394.* doi:10.2166/nh.2016.260

[26] Gao P, Carbone G J, Tufford D L, Patel A and Rounen, L F 2012 Assessing methods to disaggregate daily precipitation for hydrological simulation South Carolina Water Resources Conference: River Basin & Stream Systems Retrieved from https://tigerprints.clemson.edu/scwrc/2012/2012basin/8/

[27] Camuffo D, Becherini F and della Valle A 2020 Relationship between selected percentiles and return periods of extreme events *Acta Geophysica* **68**, 1201-1211(2020) doi:10.1007/s11600-020-00452-x

[28] USEPA 2009 Technical Guidance on Implementing the Stormwater Runoff Requirements for Federal Projects under Section 438 of the Energy Independence and Security Act *EPA 841-B-09-001, United States Environmental Protection Agency*, Washington, DC 20460

[29] Fortunato A, Oliveri E, and Mazzola M R 2014 Selection of the optimal design rainfall return period of urban drainage systems *16th Conference on Water Distribution System Analysis, WDSA 2014* **89,** 742-749

[30] Nile B K, Hassan, W H, and Esmaeel B A 2018 An evaluation of flood mitigation using a storm water management model (SWMM) in a residential area in Kerbala, Iraq *IOP Conference Series: Materials Science and Engineering* **433**(2018) *012001.* doi:10.1088/1757-899x/433/1/012001

[31] Li Y, Huang J J, Hu M, Yang H and Tanaka K 2020 Design of low impact development in the urban context considering hydrological performance and life-cycle cost *Journal of Flood Risk Management* doi:10.1111/jfr3.12625

[32] Hong J and Kim L 2016 Assessment of performances of Low Impact Development (LID) facilities with Vegetation *Ecology and Resilient Infrastructure* **3**,100-109 doi:10.17820/eri.2016.3.2.100

[33] Tobio J A S, Maniquiz-Redillas, M C and Kim L H 2015 Application of SWMM in Evaluating the Reduction Performance of Urban Runoff Treatment Systems with Varying Land Use *International Low Impact Development Conference 2015* doi:10.1061/9780784479025.00

Effect of reduction of heat using Kenaf fibrous pulverised fuel ash concrete in hydration process

N M Azzmi[1], J M Yatim[2], D Mazlan[2], Norhidayah Md Ulang[3], S Z Hashim[1] , N A S Kamarudin[2]

[1] Faculty of Architecture, Planning and Surveying, Universiti Teknologi MARA, Malaysia, 32610 Seri Iskandar, Perak, Malaysia

[2] School of Civil Engineering, Faculty of Engineering, Universiti Teknologi Malaysia, 81310 Johor Bahru, Johor, Malaysia

[3]School of Housing, Building and Planning, Universiti Sains Malaysia, 11800 Penang, Malaysia

Email: noraz026@uitm.edu.my

Abstract. Since early civilization, the toughness of concrete Pozzolanic materials, either naturally occurring or artificially produced, has long been in practice. The purpose of pozzolanic materials in the cement industry especially contributes to strengthening the concrete structure after its reaction with water. However, the increase in temperature of the cement mixture after the addition of water becomes a great concern. Which can affect the quality of the structure afterward. Therefore, this research mainly focuses on the study of the heat of hydration process in the early stages of a concrete mixture. Four types of concrete mixtures are used in this study are Ordinary Portland Cement concrete (OPC), pulverised fuel ash cement concrete (PFAC), Kenaf fibrous concrete (KFC), and Kenaf pulverised fuel ash concrete (KFPC). The temperature of all samples was recorded for 72 hours. The selection of pozzolanic material of PFA cement in the mixture was found to play an important role in controlling the heat. The result indicated that the replacement of 25% of PFA (for PFAC sample) and addition of Kenaf fiber (for KFC sample) in the concrete mixture reduced the heat of hydration by 5% and 3% respectively, compared to the conventional OPC mixture. Thus, the combination of PFA and Kenaf fiber in one concrete mixture can result in a 6% reduction of the heat of hydration. Hence, the partial replacement of cement with pulverised fuel ash and kenaf fiber is beneficial, especially for mass concrete where it can minimize the effect of thermal cracking of the structure due to excessive heat rise.

1.0 Introduction

Kenaf (*Hibiscus cannabinus L.)* is one of the crop plantations that have a great potential for multiple uses in industrial composites, fabric, paper, board, and building construction. This crop can produce fiber that can be a raw construction material through the thermochemical process. Kenaf has been a promising sustainable raw material for pulp and others. It is a multi-purpose crop and can provide the raw material for industrial and energy applications. The uniqueness of this fiber can be used by its bark about 30-40% for several high-value fiber applications [1]. In this study, what is more, the concern is the performance of fibrous concrete as a composite material in building construction. This research aims to investigate the hydration process of kenaf fibrous concrete and kenaf pulverised fuel ash concrete compared to normal concrete. The matric composition contains pozzolanic material that has been identified to reduce temperature while hydration unlikely OPC will reach the ultimate temperature because of Calcium Hydroxide along the process. Hence, the addition of Kenaf fiber and Fuel Ash cement will determine the capability of the composite in the hydration process.

Indications of kenaf being put forward as a possible option for fiberboard processing for panel products became official under the 7th Malaysian Plan in the late 1990s, and now it is being researched for potential uses under the 21st Malaysian Plan [2] The Malaysian Ministry of Plantation Industries and Commodities (MPIC) has identified six key strategic areas for the development of the kenaf industry, including increasing kenaf productivity and production, improving research, development, and commercialization, encouraging commercial kenaf production, expanding on-stream industries, and promoting the use of kenaf and kenaf products It has been described as one of the latest growth items in Malaysia, where previously unutilized crops such as hemp, jute, wool and bamboo are planted to create "new value." Kenaf is to be introduced in Pahang, Kelantan and Melaka, as well as Terengganu, Perak, and Johor. The market for kenaf fibre and core as automotive materials, bio-fibers, and pulp has significantly increased due to their high environmental benefit, energy usage, bio-compos, and paper suitability.

According to the MPIC, living room panels, wall panels, and stairs are now being made from kenaf polymer fibre obtained a well-deserved reputation, and was then a common building material for making all three different building components. Kenaf as a new additive has the potential for use in renewable, "green" CO_2-sequestration concrete made from the atmosphere, which could also potentially be a better substitute for tobacco industry uses today [3]. Unfortunately, the usage of cement in the Kenaf Fiber Concrete production phase generates more CO2. Furthermore, future studies could concentrate on mineral admixtures that will help in the need for cement while assessing their effect on the life cycle from start to finish. As observed in the research and experiments with the ingredient of kenaf, the addition of fibre results in concrete results in additional efficiency for the complex and mechanical properties. Kenaf is not only showing potential in the United States but is also is being studied as a multipurpose commodity in several other countries around the world.

Finally, since this crop is currently interesting to researchers in many other countries and countries, and in the U.S. for several years, researchers, expansion has drawn their focus. So many observers believe that kenaf appears to stick out in the eyes of the government from a host of grants and support programs such as the Malaysian Timber Board's and the National Tobacco Board's (MARDI) invested in the seed. There is approximately 35 million RM [Malaysian Ringgit] worth of Ninth Malaysia's Research and Development grants allocated to date for kenaf planting. The UN Industrial Development Organization (UNIDO) supported the mission when it was awarded a grant from the Netherlands-based Commodities Fund for the production and promotion of kenaf through field-tested surveys and pilot programs.

1.1 Natural Fibrous Properties

The studies done by [4] revealed that fibre reinforced concrete (FRC) with a steel fiber ratio of 0.5% to 2% had the best overall performance. The investigation of fibre surface treatment and volume fraction in non-woven mats with a wetting agent-infused fibre material after undergoing a four-point-bending test and proving to be 40 percent stronger in flexural fractional [5]. The matrix composites were 25 times tougher, and 20% more stable. Notably, the performance of natural fibre was on a par with other synthetic fibres. Table 1 shows the research of the study on Kenaf Fiber Reinforced Concrete (KFRC). Mechanical properties performance has been determined using different parameters. The results from all researchers indicate that the inclusion of KFRC does not improve the flexural and compression strength but showed improvement in tensile strength. KFRC increases the toughness and ductility based on the distribution of crack. The inclusion of fiber length and fraction mostly in the same quantity as shown in table 1. However, they come out with a different optimum value of fiber length and fraction. The fiber length mostly used by the researcher was between 25 and 50mm and the optimum value was 0.5% and 1%.

Table 1. Research of study on Kenaf Fiber Reinforced Concrete

References	Types	Fiber Length (mm)	Fiber Volume Fraction/Fiber Content	Aggregates	Optimum value fiber volume fraction and fiber length
[6]	KFRC	0,25,30,35	0%, 1%, 2%, 3% and 4%	Sand	
[7]	KFRC	25 mm and 50 mm	0.5%, 0.75%, 1.0%, 1.5%, and 2.0%	10mm	0.75% and 50mm
[3]	KFRC	25 to 38 mm	1.2% and 2.4%	9.5mm	1.2%
[8]	KFRC	a mixture of 20mm and 50mm	0%, 1%, 3% and 5%	10mm	1.0%
[9]	KFPC	25mm and 50mm	0.5%,0.75%,1% and 1.5%	10mm	0.75% and 25mm

The performance of biocomposites depends on the properties of the fibers used in them. However, using fibers in building materials has also some disadvantages such as low modulus elasticity, high moisture absorption, decomposition in alkaline environments or a biological attack, and variability in mechanical and physical properties [10]. Fibre from the Kenaf plant can be divided into short and long fiber measured by aspects ratio. Fibre behaviour in strengthening the composite, presently research by [3] says that kenaf fiber reinforced concrete can be used as strengthening materials for reinforced concrete (RC) beams.

Concrete is a composite that remarkably good in compression but low in tensile strength. In the last few years, because of the increasing environmental concern, the utilization of fibres from natural resources (i.e. vegetable fibres) to replace synthetic carbon/glass fibres for fibre reinforced polymer (FRP) composite application has gained popularity [11]. Natural fibres, e.g. flax, hemp, jute, coir, and sisal, are cost-effective, have high specific strength and specific stiffness and are readily available [12].

Natural fiber has characteristics of high moisture absorption that may reduce the mechanical properties of concrete. In order to corporate the natural fiber with concrete, the physical and chemical treatment of fiber is a must to overcome the adhesion of fiber-matrix in a composite [13]. The natural fibers are rich in cellulose, hemicelluloses, lignin, and pectins, all of which are hydroxy l groups; they are usually hydrophilic sources and strong polar. Alkali treatment will innovate the fiber, therefore, may overcome the interfacial bonding problem [14,15]. Thus, there are major challenges of suitability between the matrix and fiber that weakens the interface region between matrices and natural fibers [16]. In a study done by Cheung et al., [17] the interfacial bonding of natural fiber and stress transfer properties in solving environmental problems have not yet been solved to date. This study shows that the major concern is the compatibility bonding of addition materials (natural fiber) to the concrete mixture. Less research about the fiber specification and fibrous concrete performance toward the environment in long term duration [18].

1.2 Problem Statement

The composite's output is more easily described as the properties of the concrete, in combination with the reinforcing fibres. The capacity to compose as lower in elasticity, strong moisture absorption, alkaline disintegration, and susceptibility in engineering properties was distracted by natural fibres [10]. Kenaf fibres may be separated by their aspect ratio into various measurement fibres. Composite enhancement behaviour, currently researched by Elsaid et al., [3], indicates that the kenaf fibre concrete beams were comparable for increasing the material strength properties . A study recorded less long-term fibre specification and fibre concrete efficiency analysis [18]. There are few existing research on the composition of the kenaf fiber-reinforced concrete in practice for the concrete mix design and assessment of the durability in long-term performance [19].

The hydration of concrete is primarily governed by materials and mix characteristics. The heat of hydration of a cement happens when the compound is reacted with water to stabilize the hydrated state of matter and similarly to the variations of chemical compositions in other composites hence, the chemical interactions are taken into consideration. Calcium is the major ingredient in Portland cement; thus, the amount of total heat generated will be affected Concrete with a high cement content might have benefits in the beginning, but a rise in temperature due to the chemical reactions causes unwanted and unpredictable cracking [20,21,22].

2.0 Materials and method

2.1 Raw Materials

There are four types of concrete mixes were prepared in this study, which is Ordinary Portland Cement (OPC) from Tasek Corporation as normal concrete, Pulverised Fuel Ash Concrete (PFAC) from Tanjung Bin Power Plant, Pontian, Johor, Kenaf fibre with OPC was called Kenaf Fibrous Concrete (KFC) and Kenaf Pulverised Fuel Ash Concrete (KFPC) was the combination of kenaf fibre and PFA cement. Table 2 shows the concrete mix proportion of the concrete mixture. Sieve analysis for the aggregate size of 10mm was referred to ASTM C136-06 for this experimental program.

Table 2. Concrete Mix Proportions

Constituent material	Proportions KFC (kg/m^3)	Proportions KFPC (kg/m^3)
Cement	463	347.3
Fly Ash	-	115.7
Fine aggregate	800	800
Coarse aggregate	867	867
Water	250	250
Kenaf Fibre 0.5%, 0.75%, 1%, 1.5%	6,9,12,15	6,9,12,15
Super plasticiser	4.63	4.63

2.2 Kenaf Fiber Treatment

This treatment plan for this fiber using 5% of sodium hydroxide (NaOH) concentration was carry out to remove hemicellulose,pectin and lignin form the raw fiber [18,23]. The fiber were soaked in the NaOH for three hours at room temperature, then were rinsed and immersed in water for 24 hours. The fiber must be dried for 3 days as it ready to be used. The National Kenaf and Tobacco Board of Kelantan, Malaysia, is the main contributor to kenaf fibre used in this research.

2.3 Research Method

Hydration happens when concrete starting to set. In the process of release heat, cement has increased the temperature and rate of hydration. The fundamental property of hardened concrete can be achieved by this testing. A rectangular plywood box size 470 × 240 × 150 mm height with the 10mm of polystyrene as insulation that covers the box was prepared. OPC concrete mixes and KFC concrete were cast in the three-gang plastic mould . Samples for PFAC and KFPC for the second batch of testing were prepared with the same procedure. Half of the concrete mix was poured inside the mould and a thermocouple type K was inserted at the centre of the mould, then the balance of concrete has fully covered the mould. The box then was covered tightly with masking tape and initial data will be started to be measured. The thermocouple connected to the data logger and the temperature data were recorded at 10 minutes interval and was continued up to 72 hours. The testing setup as shown as Figure 1. This method was referred by present studies [22] of hydration process of concrete.

Figure 1. The setup for hydration testing

3.0 Results and Discussion

The principal objective in utilizing Fly Ash is to consider using an offensive and durable pozzolanic substance. Cement from Tanjung Bin Power Power Plant, Pontian, is class F, with a chemical composition that is about 70% compared to ASTM C618 of silica, alumina and iron oxide [24]. The substitution of 25% of the weight of cement in this study was used to observe the PFA pozzolanic reaction in the hydration process. The lower percentage of Calcium Oxide (CaO) in PFA cement caused the temperature not increased to the highest peak when the hydration happened.

Based on the particle size diagram in Figure 2, it can be concluded that PFA particles are smaller than the OPC particles. For OPC and PFA, the percentages going through a 10um sieve are 30.2% and 43.2%, respectively. The particles vary from 10 μm to 35 μm in total dimension, with the smallest being less than this. Class F-PFA in particular includes calcium from 1 to 12% of the silicon and alumina material, in the form of calcium hydrate, calcium sulphate and glass.

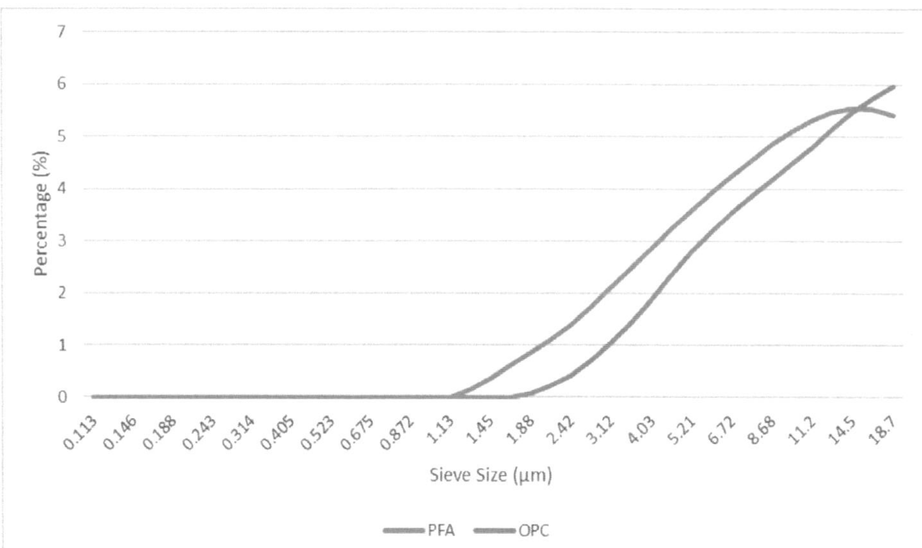

Figure 2. Particle size analysis of OPC and PFA

The C-S-H gel was found in the early stage of cement paste production to occupy around half of the volume. C-S-H sets the pore cement into a solid gel mixture. The quick hydrating mechanism somehow causes morphologies to develop tiny internal pore spaces that do not engage with the C-S-H as shown in Figure 3(a). The temperature indication of OPC, PFAC, KFC, and KFPC have been displayed in the Figure 4. As the hydration process happened, the maximum temperature of 31.6°C is reached at 16.25 hours and then drops to 25.5C. The combination of 25% Pulverised Fuel Ash Concrete in PFAC resulted in a 30.1°C maximum mixing temperature of 19.25 hours. As fibre was added , the mixture begins to set at a temperature of 30.8°C . During a mixing time of PFA and kenaf, the lowest temperature was 29.6°C, where the KFPC was reported. Previously, the addition of fiber in PFA concrete lower the cracks in the PFA and KFPC concrete sample [9]. This indicates that due to the reduction in temperature in the hydration process then less cracking occurs when the load is applied.

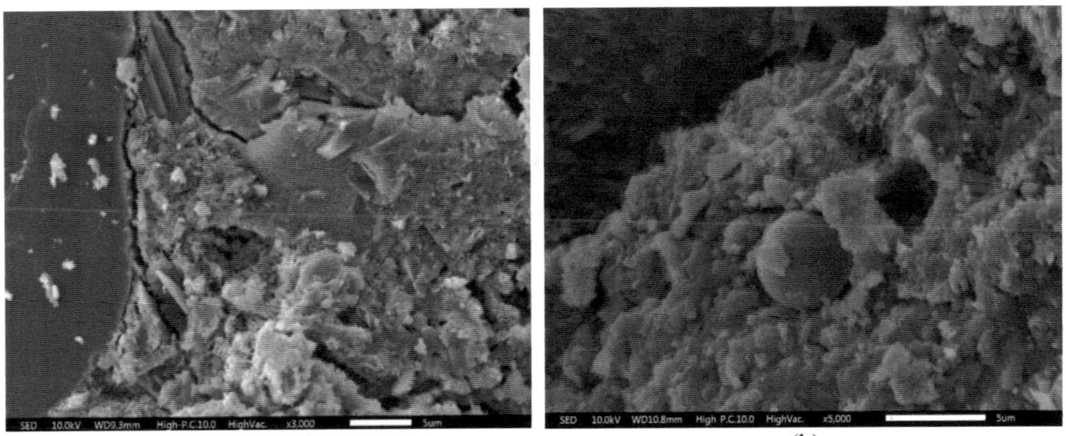

(a) (b)

Figure 3. SEM images of the microstructure of OPC and PFAC

The PFA and fibre were initiated a longer duration to bind, hence decrease the temperature, PFA to form a dense and compact matrix at a lower temperature. Fly Ash has been stated to have a little reaction in the hydration process [25]. The compressive strength of PFA was stated by Azzmi, N.M [14] indicate the compressive strength of PFA concrete was higher than OPC concrete in 56 days means it takes a longer time to enhance the strength, however it increases gradually. The greater compressive strength results demonstrated the ability of concrete to withstand a force caused by its solid matter. The hydration phase was improved by partial replacement of PFA cement as a filler in the pore structure as shown in figure 3 (b). The resistance characteristics from water penetration and other chemical attacks make it durable by providing more calcium silicates hydrate (CSH) in material composition.

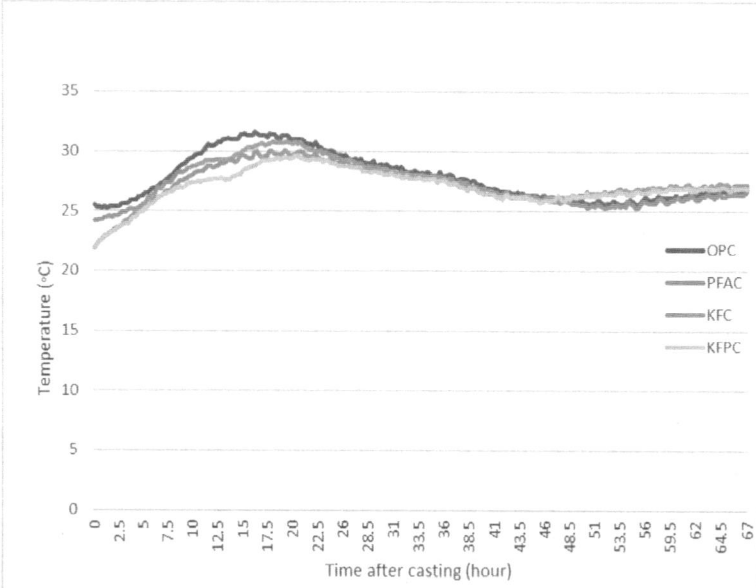

Figure 4. Development of temperature in concrete mixes

4.0Conclusion

This study shows the effect of the chemical composition of PFA cement that improves concrete to dense compared to normal concrete. The later reaction of the pozzolanic cement, has been described as a special additional concrete cementing material that enhanced the strength of concrete. The results obtained and the findings made in this study indicate that the partial substitution by pulverised fuel ash is beneficial, particularly for mass concrete where thermal cracking is of vital importance due to excessive heat increase. The current analysis was carried out with a 25% ash replacement, KFPC to minimize temperature in the hydration phase compared to normal concrete.

In general, other replacement values of PFA cement need to be considered in further study to achieve higher performance of concrete in the transition of cement paste to concrete in the hydration phase. The hydration mechanisms play important role in the mechanical properties of concrete.

References

[1] F. E. Robinson (1988). Kenaf: A new fibre crop for paper production. Calif. Agric42, 31-32.

[2] Abdul Khalil, H. P. S., Yusra, A. F. I., Bhat, A. H., & Jawaid, M. (2010). Cell wall ultrastructure, anatomy, lignin distribution, and chemical composition of Malaysian cultivated kenaf fiber. Industrial Crops and Products, 31(1), 113–121.

[3] Elsaid, A., Dawood, M., Seracino, R., & Bobko, C. (2011). Mechanical properties of Kenaf fiber reinforced concrete. Construction and Building Materials, 25(4), 1991–2001. https://doi.org/10.1016/j.conbuildmat.2010.11.052

[4] Baruah, P., & Talukdar, S. (2007). A comparative study of compressive, flexural, tensile and shear strength of concrete with fibres of different origins. *Indian Concrete Journal*, *81*(7), 17–24

[5] Li, Z., Wang, L., & Ai Wang, X. (2007). Cement Composites Reinforced with Surface Modified Coir Fibers. Journal of Composite Materials,41(12),1445–1457. https://doi.org/10.1177/0021998306068083

[6] Ahmad, P. M. I. M. A. P. M. Z., & Mahamood, P. M. A. R. (2004). The Physical and Mechanical Properties Of Concrete Reinforced With Kenaf Fiber *(Hibiscus Cannabinus L.)*.

[7] Lam, T.F. and Yatim, J.M. (2015). Mechanical properties of Kenaf Fibre reinforced concrete with different fibre content and fibre length. *Journal of Asian Concrete Federation, 1*(1), 11–21.
[8] Hasan, N. S., Sobuz, H. R., Auwalu, A. S., & Tamanna, N. (2015). Investigation into the Suitability of Kenaf Fibre to Produce Structural Concrete. *Advanced Materials Letters*, (x).
[9] Azzmi, N. M., & Yatim, J. M. (2018). Kenaf Fibrous concrete: Mechanical Properties with different fibre volume fraction. International Journal on Advanced Science, Engineering and Information Technology, 8(4), 1036–1042.
[10] Swamy R.N. (1990). Vegetable Fiber Reinfored Cement Composites - A False Dream or A Potential Reality. In H. S. Sobral (Ed.), *Second International RILEM Symposium, Vegetable Plants and their Fibres as Building Proceedi Materials, Proceeding 7* (pp. 3–8). London: Chapman and Hall, 2–6 Boundary Row.
[11] Yan, L., Chouw, N., & Yuan, X. (2012). Improving the mechanical properties of natural fibre fabric reinforced epoxy composites by alkali treatment. *Journal of Reinforced Plastics and Composites*, *31*(6), 425–437.
[12] Yan, Libo, & Chouw, N. (2013). Crashworthiness characteristics of flax fibre reinforced epoxy tubes for energy absorption application. *Materials and Design*, *51*, 629–640
[13] Cao, Y., Sakamoto, S., & Goda, K. (2007). Effects of Heat and Alkali Treatments on Mechanical Properties of Kenaf Fibers. *16Th International Conference on Composite Materials*, 1–4.
[14] Azzmi, N. M., Yatim, J. M., Hamid, H. A., Aziz, A. A., & Adole, A. M. (2018). Mechanical properties of pulverised fuel ash concrete Kenaf fibrous. *In MATEC Web of Conferences* 250 (Vol. 05007).
[15] N.Mizal Azzmi , J.M. Yatim , N. Mohd Noor , N. Md Ulang, A. M. A. (2019). Performance of Kenaf Fibrous Pulverised Fuel Ash Concrete In Acidic Environment. In *4th International Conference on Rebuilding Place*. The European Proceedings of Multidisciplinary Sciences EpMS. https://doi.org/https://doi.org/10.15405/epms.2019.12.49 ICRP
[16] A. Shalwan and B. F. Yousif, "In state of art: mechanical and tribological behaviour of polymeric composites based on natural fibres," Materials & Design,vol.48, pp.14–24,2013.
[17] Cheung, H. Y., Ho, M. P., Lau, K. T., Cardona, F., & Hui, D. (2009). Natural fibre-reinforced composites for bioengineering and environmental engineering applications. *Composites Part B: Engineering*, *40*(7), 655–663.
[18] Babatunde, O. E., Yatim, J. M., Razavi, M., Yunus, I. M., & Azzmi, N. M. (2018). Experimental Study of Kenaf Bio Fibrous Concrete Composites. *Advanced Science Letters, 24*(6), 3922-3927.
[19] Baghban, M. H., & Mahjoub, R. (2020). Natural kenaf fiber and LC3 binder for sustainable fiber-reinforced cementitious composite: A review. *Applied Sciences (Switzerland)*, *10*(1). https://doi.org/10.3390/app10010357
[20] Orchard DF (1973). Concrete Technology, Vol.1, Applied Science Publishers Ltd.
[21] Ramachandran VS, Feldman RF and Beaudoin JJ (1981). Concrete Science: Treatise on Current Research. Heyden & Son Ltd.
[22] Abdul Awal, A. S. M., & Warid Hussin, M. (2011). Effect of palm oil fuel ash in controlling heat of hydration of concrete. *Procedia Engineering*, *14*, 2650–2657. https://doi.org/10.1016/j.proeng.2011.07.333
[23] Mahjoub, R., Yatim, J. M., Mohd Sam, A. R., & Hashemi, S. H. (2014). Tensile properties of Kenaf Fibre due to various conditions of chemical fibre surface modifications. *Construction and Building Materials*, *55*, 103–113.
[24] Abubakar, A. U., & Baharudin, K. S. (2012). Properties of concrete using Tanjung Bin power plant coal bottom ash and fly ash. International Journal of Sustainable Construction Engineering & Technology, 3(2), 56–69.

[25] Deschner, F., Winnefeld, F., Lothenbach, B., Seufert, S., Schwesig, P., Dittrich, S., Neubauer, J. (2012). Hydration of Portland cement with high replacement by siliceous fly ash. *Cement and Concrete Research*, *42*(10), 1389–1400. https://doi.org/10.1016/j.cemconres.2012.06.009

Tracking Performance of Pneumatic Position Using Fractional-Order PI$^\lambda$D$^\mu$ Controller

Mohamed Naji Muftah [1,2] and Ahmad Athif Mohd Faudzi [1,3]

[1] School of Electrical Engineering, Faculty of Engineering, UTM, Skudai, Malaysia.
[2] College of Electronic Technology-Bani Walid, Libya.
[3] Centre for Artificial Intelligence and Robotics (CAIRO), UTM, Skudai, Malaysia.

E-mail: athif@utm.my

Abstract. An Intelligent Pneumatic Actuator (IPA) system is known to be highly nonlinear, subject to nonlinearities which make it difficult to control the precise position of the actuator. Thus, system modelling is crucial in creating a mathematical model that demonstrates the dynamics of a system. In this study, the pneumatic system has been modelled by an autoregressive with exogenous input (ARX) model structure, while PID and fractional orders PI$^\lambda$D$^\mu$ (FOPID) were used as control strategies. Finally, a simulation-based comparative analysis was evaluated and addressed to decide which controller provides a better tracking performance (Steady State Error (e_{ss}), Percentage of Overshoot (*%Os*), Settling Time (*Ts*), and Rise Time (*Tr*)).

1. Introduction

The intelligent pneumatic actuator chosen for this analysis has a variety of benefits, including the ability to interact and control local functions while minimizing the number of cables attached, and having more intricate, high-performance actuator motions [1], [2]. This actuator has a different trait from most of the pneumatic actuators that have full extend and contract capabilities. The pneumatic stroke in this actuator can be controlled accordingly. Unfortunately, pneumatic actuators must deal with high friction forces, dead time (due to air's compressibility), valve dead zone concerns, mass flow rate specifications, conformity variance, and producing force [3], [4]. These nonlinearities and uncertainties make modelling and precise position control of a pneumatic actuator difficult to attain [5]. Many researchers have explored the use of various control protocols in pneumatic systems thoroughly in order to address these difficulties. The pneumatic mechanism must be modelled in order to construct the controller. These models are implemented in the design control explicitly or indirectly. Overall, there are two approaches to evaluate machine mathematical modelling. The first approach is a systematic study of the law of nature. The second approach is to use the system identification techniques developed by the innovation model's experiments. Linear controllers such as PID controllers, pole-placement controllers, intelligent controllers (fuzzy and neural networks), predictive controllers, and so on were studied in the early stages of controller design [6]–[10].

In recent years, interest in Fractional Order PID (FOPID) control, the enhanced version of PID control, has grown gradually. FOPID control has a straightforward control architecture that is similar to conventional PID control but more modular, which is particularly useful for systems with nonlinear dynamics and unknown parameters [11]. In addition, the presence of two additional tune-enabled derivative and integral orders has improved the versatility of FOPID relative to PID control, in particular by minimizing residual steady-state oscillation and retaining system stability [12]. The results of these observations are that FOPID controller has been applied in a variety of programmers, including robotic systems, power generation, thermal system and permanent magnet brushless motor. However, the FOPID controller factors chosen is a potential concern because all five factors requires proper matches to attain the desired performance and consistency [13].

The main objective of this research is to create a model and design the pneumatic system controller. The pneumatic system model in the transfer function is acquired through the System Identification (SI) method. The design of the FOPID was chosen as the pneumatic system's new control technique. Controller performance evaluation was conducted with MATLAB simulation.

The paper has been organized: Section 2 experimental setup. Section 3 represents the dynamic model of the IPA. Section 4 discusses the controller design (FOPID). The simulation results to verify the system performance with the suggested control technique are introduced in section 5. Eventually, section 6 presents with some final remarks.

2. Methodology

2.1. Experimental Setup

A laser stripe code, optical sensor, pressure sensor, valves, and Programmable System on Chip (PSoC) control board are the components of the pneumatic actuator that is used in this research. The pressure sensor and on/off valves mounted to the cylinder's end are used to monitor the pressure in the chamber and monitor the cylinder's air flow, respectively. A KOGANEI-ZMAIR optical sensor identifies the location of the cylinder rod, and a laser stripe code with a smaller pitch of 0.01mm was calculates its value. The second chamber's valve is operated by a Pulse-Width Modulator (PWM) signal, and the service cycle of this PWM signal is critical for the cylinder's right and left motions.

Figure 1 illustrates the experimental setup for this analysis. Using the National Instrument (NI) Data Acquisition (DAQ) card PCI/PXI-6221 (68-Pin) desk, SHC68-68-EPM cable and SCB-68 M series modules, the pneumatic actuator system and the Personal Computer (PC) (MATLAB platform software) interact.

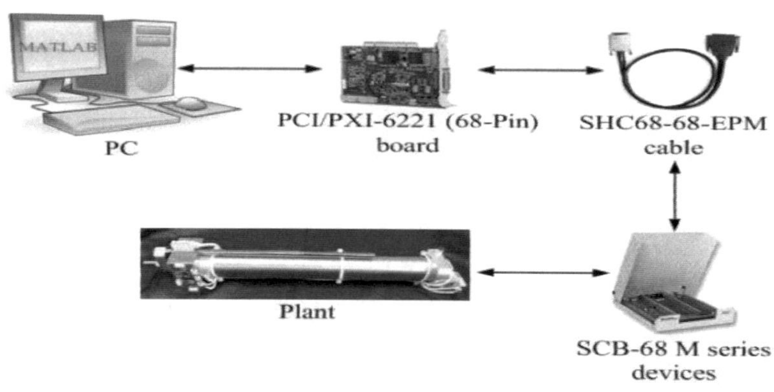

Figure 1. The experimental setup of the system.

2.2 System Modelling

A study consisting of the pneumatic actuator method described previously was performed in real time and 1600 input and output data measurements were obtained using system identification method. The first 800 samples for training while 800 samples for validation. The sampling time (Ts) utilized during collection of the input and output data is 0.01s. Figure 2 displays the plot of both loaded and obtained data.

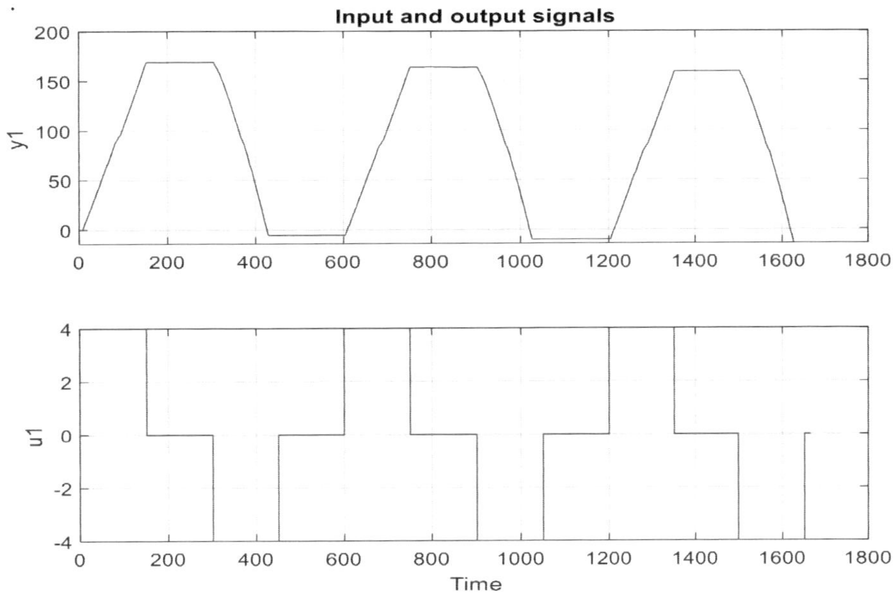

Figure 2. Input and output data of the system.

Several parametric model constructs, such as auto-regressive with exogenous input (ARX), auto-regressive moving average with exogenous input (ARMAX), box Jenkins (BJ), and output error (OE), can be used to describe the system in system identification [14]–[16]. Only ARX is used as a reference framework to reflect the structural calculation of the IPA system in this analysis. Based on its Best Fit and Akaike's Final Prediction Error (FPE), the mathematical model obtained is validated. This would be achieved by evaluating the data obtained during the simulation. The discrete ARX transfer function that used in this analysis is described in Equation (1).

$$\frac{B_{Position}\,(Z^{-1})}{A_{Position}\,(Z^{-1})} = \frac{0.1187\,Z^{-1} - 0.235\,Z^{-2} + 0.1169\,Z^{-3}}{1 - 2.99\,Z^{-1} + 2.981\,Z^{-2} - 0.9917\,Z^{-3}} \tag{1}$$

Figure 3 depicts the layout of the calculated (identified) model performance. Figure 4 depicts the zero-pole point of the established third-order ARX model.

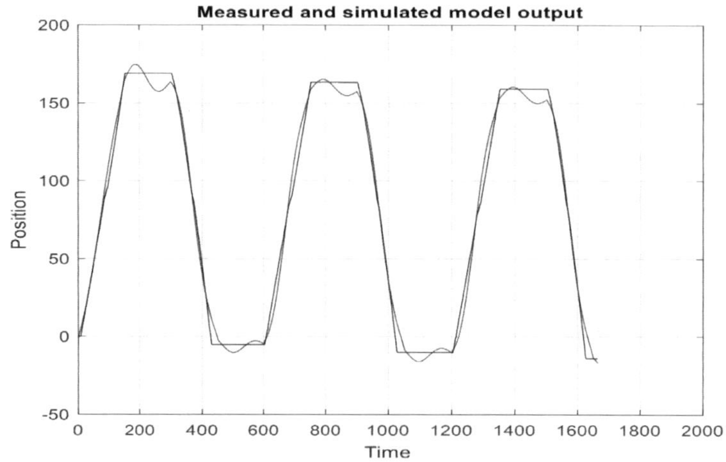

Figure 3. The measured and simulated model output

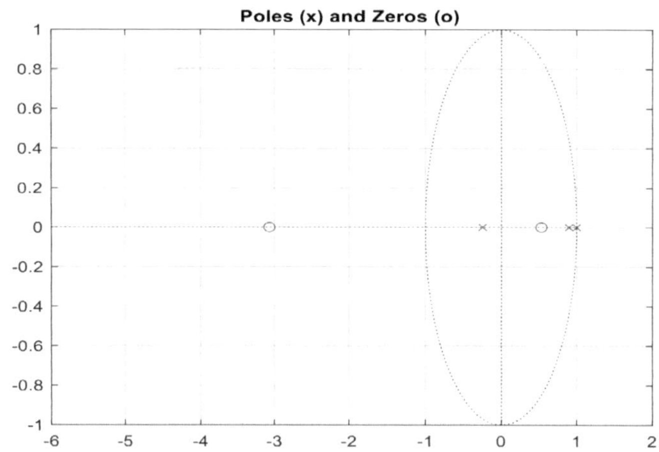

Figure 4. The zero-pole plot for the model

The model plant is valid as the compatibility is more than 90%. From Figure 3, the best fit of the output model is 90.75% whereas the Akaike's FPE of the ARX331 model is lesser at 0.04293. that means the model of IPA that we got is acceptable.

2.3 FOPID Controller

The FOPID control, as described earlier, is an expansion of the classic PID control. In the derivative and integral terms, two additional non-integer orders (λ and μ) are introduced in this control method. By using fractional calculus, the FOPID control is an efficient way to improve the durability and stability of conventional PID control [13], [17]. Equation (2) can be used to describe the FOPID control scheme.

$$C(s) = \frac{U(s)}{E(s)} = K_p + \frac{K_i}{S^{\lambda}} + K_d\, S^{\mu} \qquad (2)$$

where C(s) is control signal of FOPID controller, E(s) is errors between the relation and the real pitch position in the Laplace variable, while the K_P, KI, λ, K_D, and μ are representative of the tunable parameters of the FOPID control.

The two additional tuning parameters (λ, μ) provide additional flexibility for tuning the controller, which improves the precision and robustness of the control operation. Figure 5 illustrates the variety of these values in detail. Based on the value of integral and differentiator can know the type of controller. The controller is proportional controller when the value of (λ, μ) is (0,0), PD controller when the values are (0,1), PI controller when the values are (1,0), while PID controller when the values are (1,1).

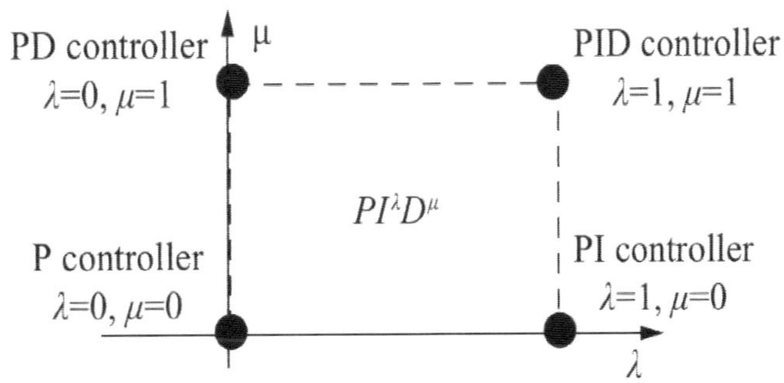

Figure 5. PID controllers with fractional-orders [18].

Figure 6 displays the IPA model with the controllers in MATLAB SIMULINK. The main goal of controller is to hold the IPA position in its optimal position.

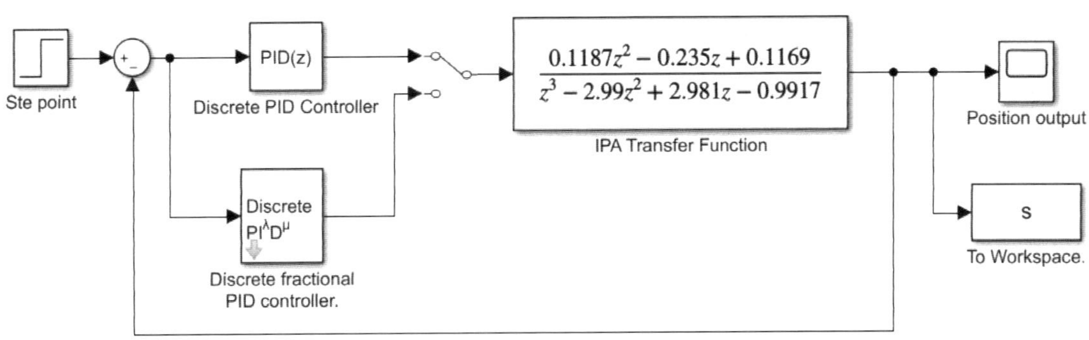

Figure 6. IPA with PID and FOPID controllers.

3 Results and Analysis

The model developed for IPA is controlled with both the classic PID and FOPID controller. The transfer function of IPA that got from system identification is described in equation (1). Table 1 includes the values of the PID and FOPID parameters that were tuned and used in this research.

Table 1. The values of PID and FOPID parameters.

Criterion	K_P	K_I	K_D	λ	μ
PID	15	1	5	-	-
FOPID	15	1	5	0.4	0.6

Table 2 displays the system performance (*Ts, Tr, OS%,* and *Tp*) in the time domain after applied PID and FOPID controllers.

Table 2. Simulation result of PID and FOPID

Criteria	Controllers	
	PID	FOPID
Rise Time, T_r (s)	0.8717	0.7992
Settling Time, T_s (s)	1.7638	0.9791
Overshoot, OS (%)	0.9150	0.0962
Peak Time, T_p (s)	3	1

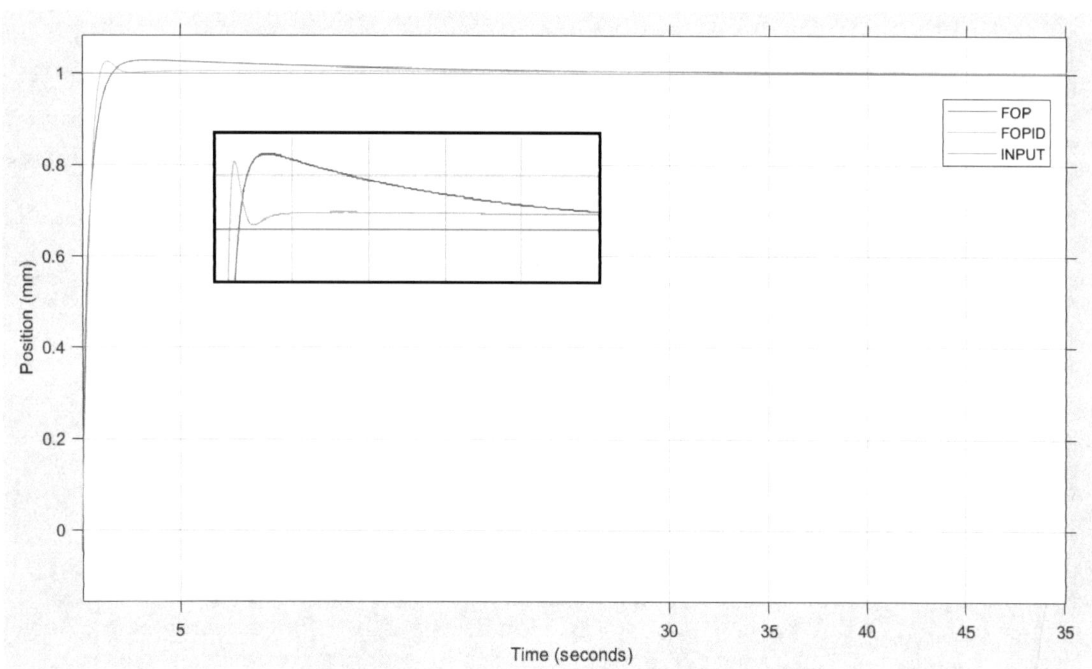

Figure 7. PID and $PI^{\lambda}D^{\mu}$ Tuned Response

The idea was to first determine the optimal parameters for the conventional PID control algorithm and then to use these optimal parameters as known parameters for FOPID control algorithm in order to determine optimal exponents of differentiation and integration (λ, μ).

Several strategies based on different control theories were recently proposed for the problem of IPA position control. Some of strategies showed an increased steady-state error and some of them recorded a longer settling time to achieve steady-state value.

In this study, The findings in Table 2 and Figure 7 reveals the FOPID is more effective and produces a more efficient response than a conventional PID controller. The implementation of fractional orders has established an accelerated response rise and settling time, enhancing the controller performance. The result shows that FOPID has 0.0962 of %OS and approximately 0 value of e_{ss} while PID has 0.9150 of %OS and also has approximately 0 value of e_{ss}. The value of Tr is 0.7992 s for FOPID and 0.8717s for PID. The value of Ts shows that FOPID gives 0.7847s faster respond compared to PID where the value of Ts is 0.9791s. By comparing both integer-order and fractional-order control time-domain performances, FOPID is found to effectively surpass the conventional method, by settling quickly to the required reference.

4 Conclusion

In this paper, the system identification toolbox is utilized in developing the intelligent pneumatic actuator model. In addition, the controller designed by using $PI^{\lambda}D^{\mu}$ Controller is used and presented. The performance of PID and FOPID are compared on their settling time (*Ts*), rise time (*Tr*), peak time (T_p), percentage of overshoot (*OS%*), and percent steady state error (e_{ss}). The simulation result found that $PI^{\lambda}D^{\mu}$ controller gives better control performance in terms of zero percentage of overshoot and percent steady state error compared of PID controller.

Acknowledgment

The authors would like to acknowledge the support provided by Ministry of Higher Education Malaysia (MOHE) through support under Fundamental Research Grant Scheme (FRGS/1/2019/TK04/UTM/02/41). The authors also would like to thank Universiti Teknologi Malaysia (UTM) for this research and facilities support.

References

[1] A. A. M. Faudzi, K. Suzumori, and S. Wakimoto, "Development of an intelligent pneumatic cylinder for distributed physical human-machine interaction," *Adv. Robot.*, vol. 23, no. 1–2, pp. 203–225, 2009, doi: 10.1163/156855308X392744.

[2] A. A. B. M. Faudzi, "Development of intelligent pneumatic actuators and their applications to physical human-machine interaction system." 岡山大学, 2010.

[3] G. Priyandoko, M. Mailah, and H. Jamaluddin, "Vehicle active suspension system using skyhook adaptive neuro active force control," *Mech. Syst. Signal Process.*, vol. 23, no. 3, pp. 855–868, 2009, doi: 10.1016/j.ymssp.2008.07.014.

[4] C. H. C. H. M. Chen, Y. P. Shyu, "Design and Realization of a Sliding Mode Control Scheme for a Pneumatic Cylinder X-Y Axles Position Servo System," *IET*, pp. 416–421, 2010.

[5] R. B. Van Varseveld and G. M. Bone, "Accurate position control of a pneumatic actuator using on/off solenoid valves," *IEEE/ASME Trans. Mechatronics*, vol. 2, no. 3, pp. 195–204, 1997, doi: 10.1109/3516.622972.

[6] N. G. Shakev and A. V. Topalov, "Continuous sliding mode control of a quadrotor," *J. Dyn. Syst. Meas. Control*, vol. 119/579, no. September 1997, pp. 441–458, 2015, doi: 10.1007/978-3-319-18290-2_21.

[7] S. Qiang and L. Fang, "Improved control of a pneumatic actuator pulsed with PWM," *Proc. 2nd IEEE/ASME Int. Conf. Mechatron. Embed. Syst. Appl. MESA 2006*, vol. 2, no. 2, pp. 1–4, 2007, doi: 10.1109/MESA.2006.297011.

[8] A. A. M. Faudzi, K. Osman, M. F. A. Rahmat, K. Suzumori, N. D. Mustafa, and M. A. Azman, "Real-time position control of intelligent pneumatic actuator (IPA) system using optical encoder and pressure sensor," *Sens. Rev.*, vol. 33, no. 4, pp. 341–351, 2013, doi: 10.1108/SR-09-2012-692.

[9] A. A. A. Emhemed, R. Bin Mamat, and A. A. M. Faudzi, "A new predictive control technique for force control of pneumatic actuator plant," *2015 10th Asian Control Conf. Emerg. Control Tech. a Sustain. World, ASCC 2015*, 2015, doi: 10.1109/ASCC.2015.7244883.

[10] K. Osman, A. A. M. Faudzi, M. F. Rahmat, N. D. Mustafa, and K. Suzumori, "Predictive functional controller design for pneumatic actuator with stiffness characteristic," *2013 IEEE/SICE Int. Symp. Syst. Integr. SII 2013*, pp. 641–646, 2013, doi: 10.1109/sii.2013.6776700.

[11] J. Zhao, J. Wang, and S. Wang, "Fractional order control to the electro-hydraulic system in insulator fatigue test device," *Mechatronics*, vol. 23, no. 7, pp. 828–839, 2013, doi: 10.1016/j.mechatronics.2013.02.002.

[12] A. Tepljakov, B. B. Alagoz, C. Yeroglu, E. Gonzalez, S. H. HosseinNia, and E. Petlenkov, "FOPID Controllers and Their Industrial Applications: A Survey of Recent Results 1," *IFAC-PapersOnLine*, vol. 51, no. 4, pp. 25–30, 2018, doi: 10.1016/j.ifacol.2018.06.014.

[13] N. M. H. Norsahperi and K. A. Danapalasingam, "Particle swarm-based and neuro-based FOPID controllers for a Twin Rotor System with improved tracking performance and energy reduction," *ISA Trans.*, vol. 102, pp. 230–244, 2020, doi: 10.1016/j.isatra.2020.03.001.

[14] S. F. Sulaiman *et al.*, "Enhanced position control for pneumatic system by applying constraints in MPC algorithm," *Int. J. Electr. Comput. Eng.*, vol. 7, no. 3, pp. 1633–1642, 2017, doi: 10.11591/ijece.v7i3.pp1633-1642.

[15] L. Ljung, "System identification toolbox," *Matlab user's Guid.*, vol. 1, p. 237, 2011.

[16] K. O. Siti Fatimah Sulaimana, M. F. Rahmat, A. A. M. Faudzi, "Linear and Nonlinear ARX Model for Intelligent Pneumatic Actuator Systems," *J. Teknol.*, vol. 6, pp. 21–28, 2016.

[17] H. S. Sánchez, F. Padula, A. Visioli, and R. Vilanova, "Tuning rules for robust FOPID controllers based on multi-objective optimization with FOPDT models," *ISA Trans.*, vol. 66, pp. 344–361, 2017, doi: 10.1016/j.isatra.2016.09.021.

[18] A. Al-Mayyahi, A. A. Aldair, and C. Chatwin, "Control of a 3-RRR Planar Parallel Robot Using Fractional Order PID Controller," *Int. J. Autom. Comput.*, vol. 17, no. 6, pp. 822–836, 2020, doi: 10.1007/s11633-020-1249-9.

IOP Conf. Series: Materials Science and Engineering 1153 (2021) 012012 doi:10.1088/1757-899X/1153/1/012012

The framework for assessing public transportation by using competitiveness index indicators

G K Sinniah[1], X Y Li[1]* and S Abdulkarim

Department of Urban and Regional Planning Faculty of Built Environment, Universiti Teknologi Malaysia, 81310 Skudai, Johor Bahru, Johor, Malaysia

E-mail: xiangyu@graduate.utm.my*

Abstract. Many institutes measured and published the competitiveness ranking, in various fields of study and research as well as in the urban planning and development processes. While the city transportation is one of the main components that always be assessed from different dimensions, however people prefer to focus on pre planning and construction by exploring and measuring the effectiveness level after completion. In this paper, we assess public transportation competitiveness in terms of the bus system based on a self-evaluation framework built from numerous measuring variables to identify the current situation in the study case city Johor Bahru, Malaysia. We comply with the PCA (Principle components analysis) to divide the evaluating indicators to three categories, and eventually use TOPSIS and Fuzzy algorithm to evaluate the bus system. The study found significant improvement compared with the previous study evaluation that the competitiveness class from Bronze to Silver ranking. Simultaneously, we gained an evaluating framework from infrastructure, available and perception aspect that could be used in other places. The result verified that it is necessary and meaningful to verify the performance of the public transit system after completion due to we could revise or optimize it immediately to reach the expecting.

1. Introduction

Not everybody can afford to own private transportation because of the cost involved when traveling from one place to another. By choosing a public transport mode, it is very cost-effective, safe, and easy to access. Besides, to the National Express Transit, busses will emit 20% less carbon monoxide than one private vehicle. The growing population and economic growth are vital for providing excellent public bus services to support the expansion of urban activities. Simultaneously, the definition of public transportation competitiveness is the ability to attract travellers to use public transportation compared to other competing travel modes under certain traffic conditions.

Malaysia is a federation comprising 13 states and three federal territories. Furthermore, among these, the infrastructure facility situation is slightly different; likewise, the people's attitude for public transportation is also different in extent. For instance, the public transportation contributed around 20% of Kuala Lumpur's passenger movement people, whereas it is about 15% in Johor Bahru [1].

Johor Bahru located in the south of Malaysia as Figure 1 shows, and is one of the four major urban centres in Malaysia. It has been identified to have the potential to become competitive cities based on a study conducted by the Economic Planning Unit (EPU) with the World Bank in 2015. According to the Johor Public Transport Masterplan (2015), the Johor Bahru District population was 1,003,000 (Department of Statistics Malaysia, 2015) and expected to have a population of 4,715,587 by 2045. The local authority devotes create public transportation to relieve the serious urban traffic stress [17].

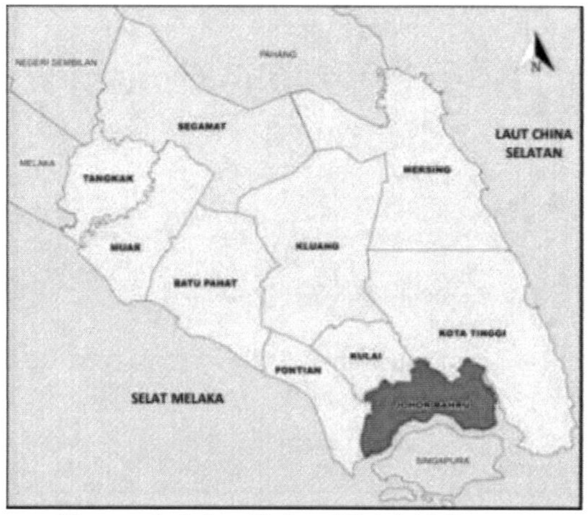

Figure 1. Johor Bahru District Map
Source: Johor Bahru Transport Mastcrplan (2015-2045)[17]

2. Problem Statement

Customer satisfaction in public transportation is vital to recognize and analyze passengers' expectations and demands [2]. It is also significant for the supplier to build and improve this system to reach the desired result. Under these backgrounds, some institutes establish a holistic evaluation system to compare the central city worldwide or in a specific realm to reveal the urban living situation or city's sustainable developing depth.

Such as the World Economic Forum published "The Global Competitiveness Report 2017–2018" that devote identify the main challenges and barriers to growth facing their economies to make economies more competitive, productive, and prosperous. The urban transport competitive ranking also formed due to the rapid urbanization in which the urbanized population in the world would increase from 55% (2015) to 68% (2050).

In Malaysia, urbanized population would rise from 75% in 2015 to 90% in 2050 (Data from the United Nations "The Prospect of World Urbanization"). Moreover, urban transport not only relieves the traffic stress because of mass transit capacity but also reduces per unit consumption of energy that benefits the natural environment and human health. According to the estimation of 'Low Carbon Society Blueprint for Iskandar Malaysia, 2025' [3], the auto ownership is expected to grow from 500 cars per 1000 people to more than 800 cars per 1000 people by 2025 because of the lack of adequate public transportation that leads to the growth of motor vehicles.

However, there is less emphasis on the public transportation factors that influence the city ranking in many competitive index reports, especially in Malaysia's realm. Malaysia is a famous tourist resort globally, thereby scholars usually research this aspect. Such as the author identified and assessed the tourist destination competitiveness by questionnaire to enable the industry stakeholders to manage and maintain the destination tourism resources more effectively [4].

Besides, a group is researching the overall urban transport in Malaysia from several aspects to reflect and summarize the realistic situation. Moreover, given some suggestions such as public transportation could reduce the emission of CO_2 and a good organized, the planned strategy could solve the travel distance and congestion and less fuel consumption [5]. Indeed, no single index can answer all questions, and there is a need for multiple indicators.

Thereby, in terms of the urban public transportation field, several dimensions should be aimed and applied to measure and acknowledge the exact situation. For instance, to assess Urban public

IOP Conf. Series: Materials Science and Engineering 1153 (2021) 012012 doi:10.1088/1757-899X/1153/1/012012

transportation competitiveness in Qinhuangdao City where located northeast of China, used fourteen indexes from public transportation infrastructure and service quality to reflect local public transportation situation [5]. Besides, the author [6] adopted fifteen indexes from three vision to evaluate sustainable transportation in Taipei, China. By this limited that many people focus on metropolis or capital city, whereas the rest city also significant be identify.

Changing focus to the Johor Bahru as a testing sample, according to previous evaluating that Johor Bahru scored a 50% or Bronze rating in Public Transportation Indicator. It estimated the public transportation performance according to accessibility, fare and payment, information technology and service performance these four variables, and the Bronze rating means "Rarely acceptable with major improvement of public transportation services to support competitiveness city".

Is there has any tangible improvement after the local government, implemented some measures to improve urban public transport? Thereby identify currently public transportation performance and endeavour to form competitiveness evaluating, it is significant, whereas the forecasting would be true which the public transport usage to decline from 15% to 10% by 2030 [1]. And it will remarkably increase urban traffic pressure compare with the current.

3. Literature Review

3.1. Global Index on Transportation

Competitiveness cities involve a very comprehensive concept and consider components such as commercial, institution, liveable city, green technology, transportation, and urban governance. As [7] mentioned, transportation is a crucial part of our daily lives and is undergoing significant transformation globally. They built 23 indicators to reflect a component of urban mobility from infrastructure expenditure commitment to public transportation affordability, thereby assessed 100cities around the world.

Also, many people build different evaluation index system to reflect researching area situation. Such as [8] built four dimensions to assess the public transportation competitiveness, and "fuzzy QFD" in order to translate the passenger demand into service quality specification adopted eight criteria to survey customer satisfaction. While according to these researching achievements, the commonality and heterogeneity gradually emerge that the evaluation indicator quantity is not small. Perhaps there is no uniform standard to assess the social phenomenon, whether for holistic city performance or one aspect of society. Everyone merely adopted suitable criteria to build an evaluation system to identify and determine the realistic status of interesting matter.

3.2. Indicators selection

As below Table 1 shows, people identify the competitiveness whether the urban transport or tourist field from various aspects by the different method. Moreover, the indicator selecting should follow the study area characteristic due to the discrepancy in the choosing place. Thereby, by the global indexes, the study will cover the following indicators suitable for evaluating public transportation in the Johor Bahru area because it is our sample area. This evaluation system is combined under five aspects: affordability, service performance, safety, infrastructure and investment.

Table 1. The reference about the competitiveness evaluating of the urban realm

Reference	Aim	index number	Measuring indicator	Methodology	Case study	Result
Sitao HU, Chuchu DAI,2012	assess Urban public transportation competitiveness	14	public transit infrastructure and service quality of public transit	AHP (Analytic hierarchy process)	Qinhuangdao, China	can meet the travel needs of residents, but the network structure is below the level
Shiau, Tzay An,2012	evaluating and prioritizing transport sustainability strategies	15	land use planning, shifting model, using clean energy	Sustainability compound index (SCI) assessment, AHP (Analytic hierarchy process)	Taipei	top five strategies should be mentioned such as improving transport accessibility for elderly and disabled persons etc.
Deveci et al. 2019	assess public bus operators based on the passenger demands	8	Frequency, Convenience, Driver behavior, Cleanliness and ergonomics etc.	Principal Component Analysis (PCA), Quality Function Deployment (QFD), fuzzy (IVIF) approach	Istanbul	cleanliness and ergonomics, reduced waiting time and travel time are the most mentioned by the customer
Sun, Hui, 2010	built a competitive evaluation index system for urban public transport, and analyze the its competitiveness	4	production factors, demand conditions, related and supporting industries and competitiveness of public transport enterprises	TOPSIS (Technique for Order Preference by Similarity to Solution)	Hebei province, China	there exists obvious differences between each city's public transport development level in Hebei, and the public transport development is quite unbalanced in Hebei.
Philip P. W. Wong, 2017	evaluate the competitiveness of tourist destination in Malaysia	4	Core resources and attractors, Supporting factors and resources, Destination management, Qualifying and amplifying determinants	regression analysis	Malaysia	not all competitiveness attributes are equal in importance, and there is a discrepancy in the different destination
Zakaria, 2015	estimated the public transportation performance in Johor Bahru	4	accessibility, fare and payment, information technology and service performance	PTCCI model (Public Transportation Competitiveness City Index)	Johor Bahru, Malaysia	Rarely acceptable with major improvement of public transportation services to support competitiveness city

3.2.1. Affordability. According to [9], the social case of public transport subsidies recognized the importance of accessible, available, and affordable transport for the people. Most studies on poverty and public transport estimated the percentage of monthly income which spent in public transports, Armstrong-Wright and Thiriez (1987) considers there is an affordability problem with public transport when more than 10% of households spend more than 15% of their income, and to solve this problem, decision-makers must focus on the policies and infrastructure for alternatives modes (walking and cycling) and improve access to public transport and spend the subsidies in policies and infrastructure rather than fares.

3.2.2. Service Performance. Service performance is the evaluation of actual service received by customers to determine the quality of service they have. Customer evaluation often depends on performance so that it can explain variance in measuring service quality as a whole. These techniques allow the critical aspects of the supplied services to be identified and customer satisfaction to be increased [10]. The dimension of measuring service performance among them are tangibility, reliability, providing service as promised, Etc.

3.2.3. Safety. There is a hierarchy in the quality of public transport, which consists of different class security is identified as a degree of safety from 3 aspects: Safety from crime/ Safety from accidents/ Perceptions of security. The impact of bus systems on road safety is particularly significant because they tend to be situated along major urban arterials. Furthermore, the author [11] and [12] created 23 variables to measure the correlation between city-level and traffic safety of 100 major urban areas in the United States. It shows that improving transportation networks, facilities, and infrastructure can indirectly decrease traffic facilities by encouraging non-driving transport modes. Remarkably, public transport safety performance could not be ignored, whatever for the city or itself.

3.2.4 Accessibility. According to [13], accessibility could be measured by the people's ability to reach goods, services, activities, or affected destinations. Many factors affect accessibility, such as mobility, quality, and affordability of transport modes, system, connectivity, and land use pattern. They describe two views for accessibility, from the origin (reachable activities), and the destination or catchment area [14]. These perspectives are related to the demand and supply side.

3.2.5. Public transport infrastructure. Of interest to this study, the author [15] put the infrastructure into two main types. Firstly, the hard infrastructure and secondly is soft infrastructure. Hard infrastructure is classified under the capital assets, while the soft infrastructure includes the framework to keep and maintain the different institutes. The soft infrastructure incorporates the system of delivery of services to the people. Moreover, lots of evidence verified that the hard infrastructure and soft infrastructure complement each other, which means the public transportation infrastructure would not be reasonable when lacking hard infrastructure or soft infrastructure.

3.2.6. Investment. Public transport requires numerous investments based on the technique used and the population density [16]. While there are no standard guidelines on how much a government should spend on public transport. Generally, a good public transport allocation, would be around 5% from their yearly budget. Moreover, the term 'investing" could be associated with the different activities, the expected target in these activities is to "employ" the money (funds) during the period seeking to enhance the investor's wealth. Whereas, in the public transportation field, it guarantees inhabitants' well-being, ensuring the public transport system building and operating normally.

4. Methodology

As above mentioned, there are many methods could be used in recognizing the competitiveness of the urban field, and we reply two stages to do that. Initially, we estimate the correlation among the index, then reply PCA (Principle Components Analysis) to classify the index based on the selecting indicator category, finally assess the indicator and obtain holistic evaluating result of the competitiveness by TOPSIS and fuzzy algorithm.

Johor Bahru has a few modes of public transport such as bus, train, and taxi. However, the most available public transport is the bus; others are considered when the bus is not available for similar trips by most public transport users in Johor Bahru [18]. In 2015, there were 11 bus operators serving throughout the cities. Due to the COVID-19 impacting, we adopted an internet survey to obtain evaluating data. The questionnaire quantity has been not less than 97 because the population in Johor Bahru is 1,003,000 and we hypothesis the most appropriate margin error of 10%. Meanwhile, the confidence level is the amount of uncertainty that can be tolerated is 95%. Finally, we gained 130 respondents by an internet survey.

4.1. To estimate the correlation among the index

To assess the competitiveness of the urban transportation that we should choose the reasonable and suitable variable based on study area characteristic. While the Pearson Correlation Coefficient Analysis is commonly means to do that to quantitative the inner correlation. The unit interval [-1,1] which -1 means there is a negative correlation between measuring index, vice the 1 means is positive, and if the value is 0 that means there is no correlation. And the correlation degree gradually weakens from -1 to 0 as well as 1 to 0.

4.2. To classify the index

As previous showing that we select five indicators to evaluate the urban public transportation competitiveness. While these variables have some potentially or directly correlation. Thereby we attempt using PCA (Principle components analysis) to classify it into different category. PCA is a commonly method when filtering the indicator to gain the dimensionality reduction. Meanwhile, it measures the loadings to classify it into different factor.

4.3. Assess the urban public transportation competitiveness

After that, to assess each indicator's performance, we adopt TOPSIS (Technique for Order Preference by Similarity to an Ideal Solution) a multi-criteria decision analysis method. It is a compensatory aggregation method that compares a set of alternatives by identifying weights for each criterion, normalizing scores for each criterion, and calculating the geometric distance between each alternative and the ideal alternative, which is the best score in each criterion. Moreover, the unit interval is [0,1], and the 1 means the best solution, vice the 0 means the worst solution. Besides, to obtain the holistic evaluation status of competitiveness, we use Fuzzy logic, a form of many-valued logic that truth values of variables may be any real number between unit interval [0,1], both inclusive. Furthermore, the ranking from 1 to 0 which 1 is the top and 0 is the last. The specific stage as follow shows:

4.3.1 Establishment of weighting values between indicators. 1) Normalization for all indicators: Hypothesis there are n independence indicators for evaluating public transportation performance. There is a need to normalize all indicators into dimensionless for effective comparison. 2) Weighting values for all individual indicators: Based on the real situation, our indicator is selected from other articles as above mentioned, therefore, it could be considered the weighting value is the same among different questions. Meanwhile, different factors have different quantity numerical questions, thus, to assess it, we could add it to get factor weighting value.

4.3.2 Analysis of the values of indicators. 1) Indicator influence level: Calculate the weighted normalized decision matrix

$$r_{ij} = \frac{x_{ij} - \mathrm{Min}(x_{ij})}{\mathrm{Max}(x_{ij}) - \mathrm{Min}(x_{ij})}, y_{ij} = r_{ij} \times w_{ij} \quad (1)$$

2) Calculate how close each evaluation object is to the best plan and the worst plan

$$D_i^+ = \sqrt{\sum_{j=1}^{m} w_i (y_j^+ - y_{ij})^2}\,, \quad D_i^- = \sqrt{\sum_{j=1}^{m} w_i (y_j^- - y_{ij})^2} \tag{2}$$

Where: how close each evaluation object is to the best (left) or worst (right) plan

3) Calculate the similarity to the worst condition

$$C_i = \frac{D_i^-}{D_i^+ + D_i^-} \tag{3}$$

Where: =1 if and only if the alternative solution has the best condition; and=0 if and only if the alternative solution has the worst condition.

4) Holistic level evaluation: Build the matrix depending on the questionnaire and normalize.

$$R = \begin{bmatrix} r_{11} & \cdots & r_{1n} \\ r_{21} & \cdots & r_{2n} \\ \vdots & \vdots & \vdots \\ r_{m1} & \cdots & r_{mn} \end{bmatrix} = (r_{ij})_{m*n}, \ \sum_{j=1}^{n} r_{ij} = 1 \tag{4}$$

Where: r_{ij} calculate is (1) equation

According to Fuzzy evaluation, the algorithm is Zadeh, as follow:

$\forall a, b \in [0,1]$, Zadeh algorithm is $(\vee)$ and $(\wedge)$, $a \vee b \rightrightarrows \max(a,b), a \wedge b \rightrightarrows min(a,b)$

The holistic level evaluation algorithm as follow:

$$B = \overline{A} \circ R \tag{5}$$

Where: is five evaluation level of indicator constitute as "Gold, Silver, Bronze, Basic, Entry", and the "o" means fuzzy matrix algorithm "Zadeh"

Normalize B and the highest value means the evaluation result that match correspondence standard.

4.3.3. Ranking the estimating result. For this assessment of public transportation competitiveness ranking, we are using BRT standards (2016) [19] developed by the Institute for Transportation and Development Policy with some modification to determine the rank and measure the performance of public transportation competitiveness (Table 2). It keeps consistent with last time that another one did it in 2015 that we use the same ranking index in order to tangibly see the changing comparing with last time whether is better or inverse.

Table 2. Evaluating standard modification based on Institute for Transportation and Development Policy

Standard	Description
Gold	Point and above is consistent in almost all respects with international best practice. This indicator is the highest level of operational performance and efficiency for providing high quality services.
Silver	Point and above includes most of the elements of best practice and likely to be cost effective with sufficient demand. These indicators achieve high operational and quality service.
Bronze	Point and above Bronze meet the definition of basic requirement public transport and consistency. These indicators have some characteristics that elevate above the Basic BRT requirement of public transportation. Some major improvement to support competitive cities.
Base	Point and above means the limited service of public transportation.
Entry	Point and above means merely has public transportation in somewhere and just can be used whereas there is no service performance can be said.

Source: Adaptation of Bus Rapid Transit Standard (2016)

5. Findings and Discussion

5.1. Data analysis

In conclusion, the highest respondent in this research studies are in the age group 26-40 years with 26.15%, and the second-highest age group is 18-25 years old. From a total of 130 respondents, more than half of the respondents consist of females with 56.2%, and 43.8% are male. While most of the respondents who answer this questionnaire are employed with the total number of respondents, 69 respondents represent 53.1%, the second-highest group with a total percentage of 36.2%, or 47 respondents are students. In terms of household income (monthly), most respondents have an income of RM1001 to RM2500 per month, representing the total percentage of 45.4%, which equals 59 respondents. Following this is income below RM1000, with the total percentage of 28.5% with the total number of 37 respondents due to the second majority of the respondent is a student, so their monthly income is not consistent and not higher than RM1000.

Initially, we measured the index correlation by Pearson Correlation Coefficient Analysis to identify the inner relation among whole measuring variables which verify the selecting indicator is about this survey topic. Moreover, the result as follow Figure 1 shows:

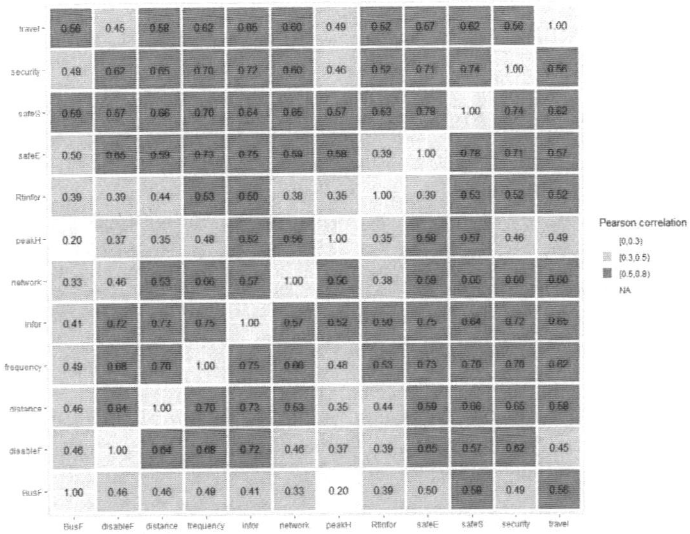

Figure 2. Pearson Correlation Coefficient Analysis

According to Figure 2 above, the significant feature could be recognized that the whole index positively correlates due to the Pearson Correlation Coefficient is big than zero. That means very indicators what we chosen are remarkable impacting with the public transportation, while the influence degree is different.

Based on this, in order to assess the public transportation from different aspect that we divided these twelve indexes into three categories by PCA (Principle components analysis), that distinguishing the indicator's inner correlation [20]. The outcome shows below Figure 3:

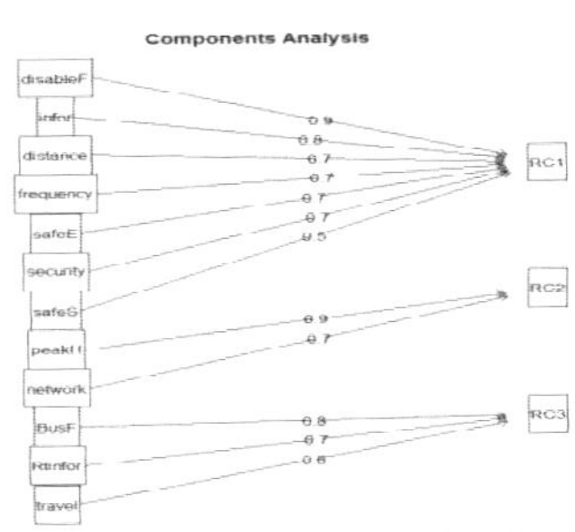

Figure 3. Components Analysis

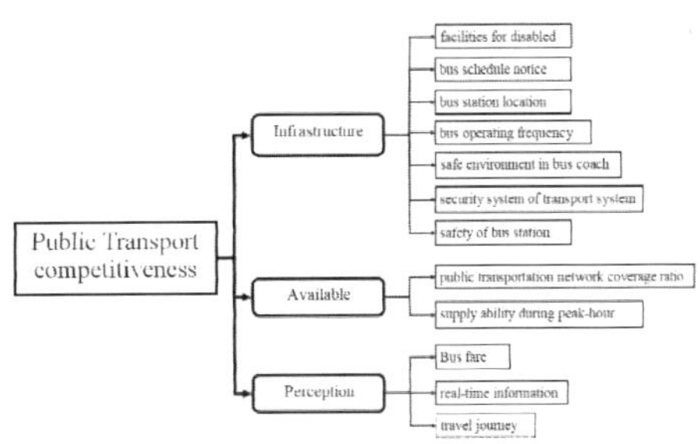

Figure 4. Evaluating framework of public transport competitiveness

These twelve indexes have been classified into three categories as the Figure 4 shows. And the correlation degree between the index and the indicator shows as the left figure, that the value higher means the correlation higher compare with other indexes what belong to the same category. Furthermore, the explanation ability as below Figure 5 shows that the Infrastructure indicator (RC1) could interpret 45% about the public transportation competitiveness as for the Proportion Explained is 0.45, which same with normal understanding due to the public transport infrastructure is the most important in this system. The Available indicator (RC2) and Perception indicator (RC3) almost same in explaining the public transport competitiveness aspect that the proportion is 28% and 27% respectively.

```
                         RC1  RC2  RC3
SS loadings             3.99 2.51 2.44
Proportion Var          0.33 0.21 0.20
Cumulative Var          0.33 0.54 0.75
Proportion Explained    0.45 0.28 0.27
Cumulative Proportion   0.45 0.73 1.00
```

Figure 5. The indicator explaining ability

Furthermore, urban infrastructure is vital for raising the overall quality of life and driving sustainable urban development, both socially and economically [21]. In terms of urban public transport, it is also significant that many urban regions enhance or consider improvements to public transportation infrastructure to address the private vehicle use challenge [22].

Our survey result shows that the public transportation infrastructure perspective obtains the most significant proportion in the whole evaluating system with 45%, the rest part Available and Perception is 28% and 27% respectively that as to the PCA calculating result. While, for the per index that we deem the weighting value is same due to all of these are positively correlation as the Pearson Correlation Coefficient Analysis measured.

Thereby, the per index measuring result as follow Table 3 shows, the best one in the evaluating framework is the Bus Fare due to its Ci value is the highest (0.72), and the following is Safa Environment in the coach because it gains 0.67 of thc Ci. While the last one is Real- time information which obtains 0.56. Actually, the all of the selecting index performance Ci value is more than 0.5 as to the unit interval is [0, 1] as for this result represents not very bad, which the 0 and 1 mean the worst condition and best condition respectively.

Table 3. The variable assessing result

Index	D+	D-	Ci
BusF	0.08355	0.215921	0.721008
safeE	0.099369	0.203503	0.671912
safeS	0.102592	0.202255	0.663463
distance	0.107077	0.203147	0.654839
travel	0.105031	0.197734	0.653094
peakH	0.107415	0.194599	0.644339
network	0.112191	0.200909	0.641677
frequency	0.11395	0.195341	0.631577
security	0.12222	0.191414	0.610309
infor	0.127152	0.188558	0.597251
disableF	0.128284	0.187789	0.594131
Rtinfor	0.136878	0.177389	0.564454

where: **disable; F**=facilities for disabled; **frequency**= bus operating frequency**; distance**= bus station location, **infor**= bus schedule notice, **safeE**= safe environment in bus coach, **safeS**= safety of bus station, **security**=security system of transport system, **network**=public transportation network coverage ratio**, peakH**= supply ability during peak-hour, **travel**= travel journey, **Rtinfor**= real-time information, **BusF**= Bus fare.

Moreover, according to the Fuzzy algorithm, we evaluated the public transportation competitiveness as below Table 4 representing:

Table 4. The holistic assessment result of public transportation competitiveness

Category		Weighting	**Bad** **Strongly Disagree**	**Basic** **Disagree**	**Bronze** **Normal**	**Silver** **Agree**	**Gold** **Strongly Agree**
Infrastructure	disableF frequency distance infor safeE safeS security	45.00%	3.85%	10.44%	19.56%	47.03%	19.12%
Accesibility	network peakH	28.00%	2.69%	9.62%	19.23%	50.00%	18.46%
Perception	travel Rtinfor BusF	27.00%	3.33%	9.23%	18.21%	51.28%	17.95%
	B		**0.0385**	**0.1044**	**0.1956**	**0.4500**	**0.1912**
B (Normalization)			**0.00**	**0.16**	**0.38**	**1.00**	**0.37**

It shows the public transportation in the Johor Bahru obtained Silver in this survey because this item gained the highest value compared with others, which our question is "Do you agree the public transportation in Johor Bahru is good?". Moreover, match it into the evaluating standard that means "Point and above includes most of the elements of best practice and likely to be cost effective with sufficient demand. These indicators achieve high operational and quality service".

5.2. Finding discussion

Our survey result shows that the public transportation infrastructure perspective obtains the most significant proportion in the whole evaluating system with 58.31%. Moreover, the respondent deemed this aspect is not bad because it obtained 47.03%, which means almost half of people agree it is good, and gain "silver" matching the evaluating standard. Besides, in this part, the safe environment and safe bus station obtained first and second grading that is 0.6719 and 0.6635, respectively, while the security facility aspect only has 0.6103. That could not be ignored because the security facility performance would be determined by IoT (Internet of Things) in extent.

This result could remarkably acknowledge the IoT facility is lagging, as to the appraisal of security facility, information notice, and real-time information system are not very good. Meantime, the disabled facility should be mentioned due to it only gets 0.5941, which means it is not bad or good following the evaluating range between 0 to 1, but it is ranked penultimate in the holistic evaluating framework.

Indeed, to reduce the of using automobility, supplied transit service must be available to as many users as possible [22]. Our evaluating framework is reflected by public transportation network coverage ratio and supply ability during peak-hour. Furthermore, in this part, 50% of people consider it good to obtain "silver" assessment, which the network grading is 0.6417, and the peak-hour performance is 0.6443. That meaning the bus network coverage ratio and peak-hour capacity enough most of the rider expectation.

To recognize it tangibly, we collected other data about bus service network situations, as the below Figure 6 shows. We hypothesize the bus station coverage range is 400 meters which the bus coverage ratio is 51.48%, and the method is total the bus station coverage area divided the construction area in

Johor Bahru. Then, we put it into the standard of "Transit Capacity and Quality of Service Manual" (United States), this coverage ratio means "bus service covers the most area in the high-density community, but the first or last mile is far and difficult to handle whether by walking and cycling". It differs slightly from the respondent's feedback.

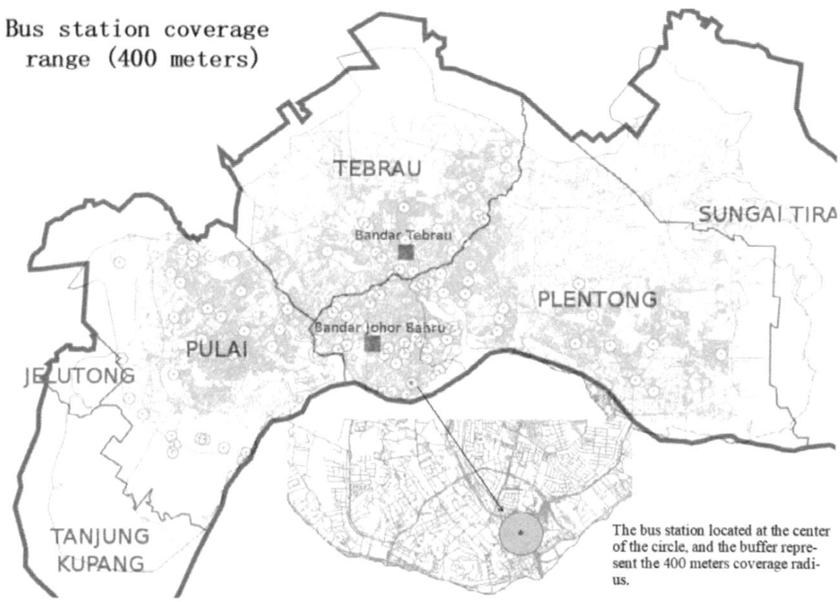

Figure 6. Bus station coverage range within 400 meters

Although the respondent deems the network coverage ratio reasonable, it also displays that accessibility should be enhanced by improving the public transportation network and increasing availability, which merely has 49 (46.7%) and 50 (47.6%) responders. Therefore, increasing bus line and bus service in the uncovered area is significant for the potential rider. It would give a positive impression hence attract more people using it to reach the goal for the supply side.

Attracting more people away from private vehicles and using public transportation is not easy. People always calculate the expenditure, mind the user experience, Etc. whatever for public transport or other things which want to obtain people mention. Wherefore, customer perception also should be cared for when assessing public transportation competitiveness.

Indeed, the perception includes subjective and objective two aspects. Our survey applied one subjective (travel journey satisfaction) and two objective indexes (real-time information and bus fare) to evaluate it. In this aspect, it gains the most considerable grading, in which 51.28% of people deemed it suitable, and the evaluating standard is "silver". Although the real-time information (0.5645) is the lowest in the whole indicator, the bus fare index (0.721) obtains the highest among it. Directly reflect the bus fare policy is successful because the resident only needs RM 10 to register the smart card and take some bus line free during the next one-year period.

However, it also clearly shows people mind the real-information system of the bus. Therefore, the agency company or department should invest more finance in the bus operating soft infrastructure to improve the public transportation infrastructure and the rider's perspective for the public transportation system.

6. Conclusion and Recommendation

Based on these studies, this paper found that Johor Bahru public transportation achieves absolute improvements compared with previous assessing. This study found out that Johor Bahru's public transportation competitive index is rated as Silver Rated, which means most of travel demand in term of public transportation could be meet by currently system. However, there still are something should be mentioned and improved to achieve the goal that reduce the private mode's share in trip marking from 75% (2005) to 35% (2035) and increase the public mode's share from 15% (2005) to 40% (28% bus,12% rail2025) [3].

6.1. Enhance the public transportation network

As above showed the bus station coverage ratio only has 51.84% if hypothesis the station covering range is 400 meters. That standard means the bus service only meet the high-density community and the first-last mile could not be solved. And it has been represented by the questionnaire survey which the respondent considered the bus availability is a shortage. While, to solve this issue that the bus operating agency could build more bus line in the vacant place as well as building the bus station. Due to the people walking willing barely has around 400-800 meters in Malaysia [23]. Thereby, it is useful to attract people from automobility to public transportation when the availability has been enhanced.

6.2. Improve the information system of the bus

The survey result shows the real-information and bus schedule notice item gain first and third from last respectively in all of assessing index. That means the people very caring about it due to them can arrange their schedule if there is reliable real-information of the bus operating. While to achieve this purpose that could improve the IoT (Internet of Things) that people could know the bus specifically operating situation in the bus station or in their smartphone. Meantime, the IoT also could optimize the security of the bus system as well as the safety of the bus station. That could significantly enhance the public transportation system from several aspects by cost-efficient means.

6.3. Reinforce publicity

Actually, to increase the public share that local authority released the discount card which people only need spend RM 10 (Adult) or RM 5 (Child & Older) to register it and could take the bus free in one year. However, only are 48(36.9%) responders done that. It clearly shows that there's a significant lack of awareness among the respondents with this initiative that has been introduced by the bus operator. Hence, the operating company and the department could publicity it via various media outlets including but not limited Facebook, Instagram and radio as well as notice it in the bus station noticeboard. Indeed, we not saw any publicity in the bus station when we waiting the bus.

As for this public transportation assessing framework, we gained some effective and meaningful outcome in the case area that could help us to identify the current public transportation competitiveness and compare the achievement after this system improving project has been implemented that done it in 2015. However, this primary evaluation focuses on survey data that perhaps has some deviation because sampling selection is holistic, and the responder is probably living an excellent bus service community. Therefore, there also are some issues with it. Hence, the evaluation of this could adopt multidimensional data to enhance result accuracy and get more objectively result.

References

[1] Zaman M, Sultan Z, Fard M, Siyaka A, Pung JC. An Assessment of Public Transport Facility in Johor Bahru: a case study in Taman Ungku Tun Aminah Area, Majlis Perbandaran Johor Bahru Tengah, Malaysia. *International Journal of Built Environment and Sustainability* 2017;**4**:71–80. https://doi.org/10.11113/ijbes.v4.n2.178.

[2] Wong PPW. Competitiveness of Malaysian destinations and its influence on destination loyalty. Anatolia 2017;28:250–62. https://doi.org/10.1080/13032917.2017.1315825.

[3] Centre U-LCAR. *Low Carbon Society Blueprint for Iskandar Malaysia 2025 - Summary for Policymakers*. vol. **53.** 2013. https://doi.org/10.1017/CBO9781107415324.004.

[4] Almselati ASI, Rahmat RAOK, Jaafar O. An overview of urban transport in Malaysia. *Social Sciences* 2011;**6**:24–33. https://doi.org/10.3923/sscience.2011.24.33.
[5] Sitao HU, Chuchu DAI and SC. *Urban Public Transport Competitiveness Evaluation Based on AHP and Extension Theory* 2012:2433–42. https://doi.org/10.1061/9780784412442.071.
[6] Tzay An S. Evaluating sustainable transport strategies with incomplete information for Taipei City. *Transportation Research Part D: Transport and Environment* 2012;**17**:427–32. https://doi.org/10.1016/j.trd.2012.05.002.
[7] Sun H, Fan Z, Shi Y. Research on comprehensive evaluation of urban public transport competitiveness in hebei province based on TOPSIS. 2010 *International Conference on Logistics Systems and Intelligent Management,* ICLSIM 2010 2010;1:250–4. https://doi.org/10.1109/ICLSIM.2010.5461426.
[8] Deveci M, Öner SC, Canıtez F, Öner M. Evaluation of service quality in public bus transportation using interval-valued intuitionistic fuzzy QFD methodology. *Research in Transportation Business and Management* 2019;**33**:100387. https://doi.org/10.1016/j.rtbm.2019.100387.
[9] Serebrisky T, Gómez-Lobo A, Estupiñán N, Muñoz-Raskin R. Affordability and subsidies in public urban transport: What do we mean, what can be done? *Transport Reviews* 2009;29:715–39. https://doi.org/10.1080/01441640902786415.
[10] Vergel-Tovar CE, Rodriguez DA. The ridership performance of the built environment for BRT systems: Evidence from Latin America. *Journal of Transport Geography* 2018**;73**:172–84. https://doi.org/10.1016/j.jtrangeo.2018.06.018.
[11] Najaf P, Thill JC, Zhang W, Fields MG. City-level urban form and traffic safety: A structural equation modeling analysis of direct and indirect effects. *Journal of Transport Geography* 2018;**69**:257–70. https://doi.org/10.1016/j.jtrangeo.2018.05.003.
[12] Laaly S. New definition of Transit Oriented Development (TOD) based on context sensitive paradigm. *ProQuest Dissertations and Theses* 2014:**171.**
[13] van Maarseveen M, Martinez J, Flacke J, Singh YJ, Flacke J, Zuidgeest M, et al. Planning for Transit Oriented Development (TOD) Using a TOD Index. *GIS in Sustainable Urban Planning and Management* 2019:267–82. https://doi.org/10.1201/9781315146638-15.
[14] Skorobogatova O, Kuzmina-Merlino I. Transport Infrastructure Development Performance. *Procedia Engineering* 2017;**178**:319–29. https://doi.org/10.1016/j.proeng.2017.01.056.
[15] Boujelbene Y, Derbel A. The Performance Analysis of Public Transport Operators in Tunisia Using AHP Method. *Procedia Computer Science* 2015;**73**:498–508. https://doi.org/10.1016/j.procs.2015.12.039.
[16] FANG W. *Comparison and Application of Principal Component Analysis & Fctory Analysis* 2003:6–8. https://doi.org/10.16309/j.cnki.issn.1007-1776.2003.03.004.
[17] Johor PPA. Johor Bahru Transport Masterplan(2015-2045). vol. 53. 2016.
[18] Moghaddasi A. Transport Cost Analysis of City Bus and Private Car Usage in Johor Bahru , *Jurnal Teknologi* 2013;**3**:25–31.
[19] Institute for Transportation & Development Policy (ITDP). the Brt Standard Introduction Brt Awards Showcase Scoring in Detail Application To. *The BRT Standard* 2016.
[20] Chen Y, Shen L, Zhang Y, Li H, Ren Y. Sustainability based perspective on the utilization efficiency of urban infrastructure — A China study. *Habitat International* 2019;**93**:102050. https://doi.org/10.1016/j.habitatint.2019.102050.
[21] Chakour V, Eluru N. Examining the influence of stop level infrastructure and built environment on bus ridership in *Montreal. Journal of Transport Geography* 2016;**51**:205–17. https://doi.org/10.1016/j.jtrangeo.2016.01.007.
[22] Eboli L, Forciniti C, Mazzulla G. Service Coverage Factors Affecting Bus Transit System Availability. *Procedia - Social and Behavioral Sciences* 2014;**111**:984–93. https://doi.org/10.1016/j.sbspro.2014.01.133.

IOP Conf. Series: Materials Science and Engineering 1153 (2021) 012012 doi:10.1088/1757-899X/1153/1/012012

[23] Azmi DI, Karim HA, Amin MZM. Comparing the Walking Behaviour between Urban and Rural Residents. Procedia - Social and Behavioral Sciences 2012;68:406–16. https://doi.org/10.1016/j.sbspro.2012.12.237.

Review on current advances and consideration issues in preparing soil-rock mixture for laboratory testing

N Salimun[1], H Mohamad[2]

[1,2] Universiti Teknologi PETRONAS, 32610, Seri Iskandar, Perak, Malaysia
Email: norsyafiqahsalimun@gmail.com

Abstract. An extensive review of the soil-rock mixture strength measurement has been carried under various geomaterials and experimental methods. Primarily, the factors affecting the SRM mechanical behaviour and the testing methods discussed to further understand the importance of the sample preparation process before conducting laboratory testing. The soil-rock mixture is prepared in the laboratory using the test scheme proposed from very recent literature to perceive the dependence of the sample preparation procedure on the result of strength measurement. Then, a thorough literature review conducted to highlight the significant observations that are important in developing a systematic procedure for soil-rock mixture laboratory testing. Furthermore, the findings conclude that besides the soil-rock ratio, the moisture content, type of soil or finer grain component, and variation block strength or coarse grain component are essential factors that need to be determined while preparing soil-rock mixture samples before conducting the laboratory testing. Finally, for the future research direction of preparing a systematic standard for soil-rock mixture samples, several key points include the particle size, the range of soil-rock ratio, the geomaterials used, and the compaction degree proposed to be considered.

1. Introduction

Soil-rock mixture (SRM) is a significant geomaterial that is widely distributed and abundant in nature. SRM described as distinctively chaotic geomaterials composed of various block rocks of differing lithologies and sizes pervasively embedded within a weaker fine-grained matrix [1]. Indifference to block-in-matrix (bimrocks), a mixture of rocks composed of geotechnically significant blocks within a bonded matrix, the matrix of SRM is looser [2]. Soil-rock mixtures are usually heterogeneous, and their strength parameters correlated to structural components such as rock content, size, block shape, nature of the soil, and spatial distribution [3], [4]. Figure 1 shows an example of a soil-rock mixture in nature.

Due to the non-linearity of the parameters and deformation characteristics of SRM, its mechanical properties measurement is more complicated than other geomaterials [5]. At the same time, there is a complex interaction between soil and rock blocks [6]. In many construction projects such as tunnels, dams, and slopes, the soil-rock mixture is common, which leads to challenging instability problems and project delays during the design and construction process. Thus, identifying the strength parameters of this kind of anisotropy geomaterials is crucial in engineering work.

Figure 1. Soil-rock mixture in nature

Currently, SRM is not separated from the traditional classification system of geotechnical materials. Even though researchers worldwide have conducted various experimental studies using SRM, a systematic standard of experimental or laboratory work is not well established. The sample preparation of the SRM used for laboratory work is complex due to the disturbance condition, sample size, and state. It is not easy to provide an identical sample between the specimens from the field and the fabricated or moulded sample, which usually used in laboratory testing.

To overcome these disadvantages, certain researchers have started to research the strength of the SRM using the numerical modelling method. Yet, most of the numerical modelling software requires the implementation of a specific constitutive model. This method can affect one's judgment of the failure mechanism of soil-rock. Therefore, laboratory model experiments are still an efficient way to understand the properties of the SRM. Yet, there are a variety of methods used in preparing the SRM sample for laboratory testing. Thus, this paper aimed to review the significant work in SRM's properties measurement using a physical model, highlight the fundamental factors that affect the SRM preparation process, and suggest a way forward to have a systematic standard on preparing the sample for laboratory work.

2. Soil-rock mixture strength measurement

Strength is among the key parameters to characterize geomaterial mechanical behaviour. Hence, various methodologies applied by researchers to measure the strength of the soil-rock mixture. Most of the studies recognized that the soil-rock ratio mainly influences the strength. Based on the popularity table constructed from the literature review (refer to Table 1), the soil-rock ratio is indeed the most popular factor studied, followed by the rock grain size. It is observed from the literature study that most works focus on the influences of gravel content and SRM's failure mechanism [4], [7], [8], as well as the influence of gravel size and gravel shape [9]–[12]. Other factors such as rock-block shape, strength, and orientation not included here as the works are minimal.

Table 1. Popularity table on previous research regarding factors affecting the shear strength of soil-rock mixture.

Paper	Soil-rock ratio	Particle size distribution	Moisture content	Rock grain size effect	Material composition
B. Li [9]	1	1	1		
Shaorui S. et al. [13]	1			1	1
Wang S, Ji T, Xue Q, et al.[8]				1	
Minghui R. et al. [10]					
Yanxi Z. & Zhongzian L. [3]				1	1
Cen D et al [14]	1			1	
Zong-Liang Z. et al. [15]					1
Hai-Yang Zhang et al. [7]		1			
Xue et al. [16]			1		
Gao W. et al. [17]	1				
Shaorui Sun et al [18]	1	1		1	1
Linping Y. & Zhiyun W. [19]					
Xu Wen-jie et.al [20]	1				
Wickland BE et al. [21]	1				
Vallejo LE [22]	1				
Popularity	8	3	2	5	4

The majority of studies measure the mechanical properties of SRM using large direct shear tests either in the field or laboratory [9], [10], [13], [16], [20], [23]–[25]. Though there are many limitations to direct shear testing, including the non-uniformity of the stresses and strains within the box, due to its simplicity and suitability for testing, the direct shear box is one of the most widely used equipments for obtaining shear strength parameters (cohesive force and internal friction angle) for geotechnical material. For instance, a shear test conducted using fabricated SRM made of breccia and subrounded gravels mixed with silty clay as fine soil [13]. The result shows that when the gravel proportion is too high (for example, 70%), the shear stress value depends on the gravel shape and the distribution in space and not completely on the gravel proportion. B. Li similarly uses a large-scale indoor shear test on SRM to investigate the effect on gradation and maximum particle size [9]. In his study, the SRM sample made of sandy gravel and quartzite. The result showed that the effect of gradation on τ, φ, and c is small. Meanwhile, as the maximum particle size increases, the τ, and φ of the SRM increase, while the c decreases.

The triaxial test is another typical method selected to study the strength of the soil-rock mixture. J. Yang et al. also used triaxial compression tests but using different moulded SRM samples made of cement sand-gravel [26]. Ren et al. also studied the specimen size effect on the mixture of clayey soil, gravel, and rock blocks [10]. They determined that the shear gap controls the shear behaviour of SRM. With the gap increment, the relationship between internal friction angle and cohesion is in negative correlation. Stable and realistic shear strength of SRM obtained in suitable gaps, and the specimen size effect also exists with the gap effect.

Other than the conventional testing method (shear and triaxial test) mentioned above, several attempts using different strategies and equipment to measure the strength of SRM presented here. For example, Lv and Zhou instead conducted a laboratory test using a modified borehole pressure shear test on the mixture of sand, calcium carbonate, gypsum [27]. They found that the strength parameters are slightly higher than the direct shear test, but this suggested alternative method considered efficient enough. Zhenping Z. et al. and Hai-Yang Z. et al. conducted a uniaxial compression test combined with computed tomography triaxial test, another different testing method of SRM sample from colluvium slope [4],[7]. Their finding introduces another two controlling factors of soil-rock mixtures' mechanical properties: interlocking and the breakage of the large rock blocks.

There are also several studies of SRM made using synthetic material. For instance, Ding et al. mould a transparent synthetic soil using a mixture of fused quartz sand with liquid paraffin and n-tridecane [23]. The results show that transparent soil can be used for similar simulation experiments of the soil-rock mixture since both of them have high similarity. Besides, Wang et al. analyze the shear properties of clayey soil with corundum ball using an electro-hydraulic servo and determined the axial and lateral deformations [28]. The experimental results define three failure patterns of SRM; spitting failure, shear failure, and conical failure at various strain rates. They also found that the interactions between the rock blocks and the soil matrix are the primary factor determining the dynamic response of SRM.

Rather than using testing equipment as mentioned above, there is also another study determined that the sand-gravel mixtures' shear strength using empirical equation $S_c = S_m\ (1+2.5\ C)$, which is developed by Vallejo et al. The equation used value from the shear strength of the sand matrix and the concentration by volume of the gravel in the mixtures. In this equation, Sc, $S_{m,}$ and C represent the shear strength of the sand-gravel mixture, the shear strength of the sand matrix alone, and the concentration by volume of the gravel in the mixture. However, the equation's validity has not yet been determined to be general and has only applied to the type and size of the grain, stress conditions, and equipment used in the reported testing program [24].

To conclude, the strength properties of the soil-rock mixture depend not only on soil-rock proportion but also on other factors, such as the rock block size, texture, shape, moisture content and the experimental conditions. Concurrently, the documentation of the strength parameters for the SRM is necessary for engineering work and could be done by utilizing the results of the model experiments that consider many influencing factors [13].

3. Soil-rock mixture sample preparation method

Variety in soil-rock mixture testing influences the sample preparation method too. Generally, soil-rock mixture (SRM) preparation for laboratory testing divided into three-phase. The first phase starts with collecting and preparing the soil-rock mixture component. The type of materials used can be collected from natural sources or fabricated. Several types of geomaterials used to prepare the soil-rock mixture sample from previous studies, as presented in Table 2. It is essential to mention here that the works of literature presented in this section only involve SRM behaviour measurement using physical testing.

Table 2. Material list used in preparing soil-rock mixture from literature

Paper	Material
Lv & Zhou [27]	Sand, calcium carbonate, gypsum
Ren et al. [10]	Clayey matrix, gravel, and rock blocks
B. Li [9]	Sandy gravel and quartzite
Xue et al. [16]	Pebbles, silty clay, and water
Ding et al. [23]	Synthetic transparent soil
Wang et al. [28]	Clay soil and corundum ball
Sun et al. [13]	Silty soil and gravel
Zhang Z. et al. [4]	Colluvium
Zhang H. Y. et al. [7]	Colluvium
Xu et al. [20]	Clay, cobbles, and gravels from SRM slope
Zhao & Liu [3]	Clayey soil and artificial rock
Vallejo et al. [24]	Sand and gravel
Zhang Z. L. et al. [15]	Residual soil and granite
Younes Amini et al. [25]	Soil with fine sand and gravel
Hamidi et al. [29]	Sandy soil with gravel
Selvam & Chakravaty [30]	Sandy silt

Qi et al. [31]	Frozen talus
Huang et al. [32]	Clay-gravel mixtures
J. Yang et al. [26]	Cement-sand-gravel (CSG)
H Yang et al. [33]	Soil and gravel

In the second phase, the procedure proceeds with designing the material composition or soil-rock ratio. A recent study conducted on the material composition effects on the mechanical properties of SRM, where the soil-rock material was separated into different size categories before tested using medium-scale shearing and triaxial experiments [3]. The relationships among macro deformation, strength, content, size, and random location of rocks observed where the stress-strain curve of soil-rock mixtures found as an approximately hyperbolic hardening curve. The SRM sample collected from the site undergoes a sieving process to separate the soil and rock particles. A study on the effect of SRM particle size on the shear strength found out that, when the Mohr-Coulomb criterion used to depict the curve under a shear strain of 0.15, cohesion first increased and decreased. It also determines that the elastic modulus increased with an increase in rock size, but Poisson's ratio remained constant.

The third phase is the moulding process. Depending on the scale and testing equipment used, the sample prepared come in different form and size. Based on previous works included in this review, the sample is usually moulded into a core sample or compacted into the shear box. Large-scale equipment can run a test with a minimum square dimension of 10cmx10cm. Meanwhile, the core sample used in the triaxial or uniaxial compressive test is in cylinder form with a minimum dimension of 100cmx50cm. A standard dimension of the sample preparation is not fixed yet in any published work.

4. Perspective and Future Direction

Based on the discussion in the previous section, various methods are existing in testing and preparing the soil-rock mixture. In this study, a soil-rock mixture core sample had been prepared by following the test procedure present in previous work using natural samples collected at a construction site from Kinta Valley. After attempting to prepare the soil-rock mixture, there are several findings can be concluded. There are three significant factors observed through the process that is going to affect the sample preparation, which discussed as the following:

4.1. Moisture content

The moisture content affects the shear strength of the soil-rock mixture. A previous study shows that the shear strength of SRM reaches its peak when the moisture content is 10.5% [9]. Consequently, the moisture content needs to be determined before preparing the laboratory sample. The moisture content of the sample is also dependent on the type of soil present in the mixture. Sandy and silty soil hold moisture differently as they have more pores compared to clayey soil. In the water film theory, too high or too low moisture content is not conductive to closely arranged granular materials as the grain squeezes each other and adversely affects the mixture's strength [34].

Besides, it is observed that the time gap between sample preparation and sample testing also affect the moisture content, consequently the testing result. The core samples are purposely left in the open air and closed box for different time ranges before running the uniaxial compressive test. The finding shows that the longer the sample kept before the testing, the lower the moisture content, the harder the sample becomes. Thus, a standard time gap needs to be established in the systematic procedure of soil-rock mixture testing to obtain more comparable data to be analyzed.

4.2. Type of soil

The soil in the soil-rock mixtures is a relative concept and different from the traditional notion of silt, clay, and other fine-grained soil. Since particle size varies from several millimetres to tens of centimetres, defining the soil and rock is vital in determining the rock content of this medium [3]. In the particle size distribution (PSD), the soil type classified as clayey, silty, or sandy. The SRM sample prepared in this study divided into two types according to the soil used. The first sample prepared using sandy soil, and the other is silty soil. The core using sandy soil is fragile compared to silty soil. Thus, the PSD analysis is a crucial procedure proposed to include in the systematic procedure of SRM laboratory testing.

4.3. Rock-block properties

The soil-rock mixture is a cohesive-frictional geomaterial subjected to impacts of composition and structure seriously [8]. The coarse-grained component, known as rock block in the soil-rock mixture, can be weathered, fabricated, or fresh. The texture of the rock block also affects the mechanical behaviour of SRM. For example, a recent study on soil-breccia mixture (a type of SRM) shows that, as the confining pressure increased, the shear strength of the soil-rock mixture increased. The amount of displacement of the upper breccias is the largest.

Usually, the sample is taken from the field and classified into a disturbed or undisturbed sample. The sample also could be made of synthetic or artificial material. Most researchers prepared the SRM sample by moulding them using either natural or synthetic material, as shown in Table 2. Some other researchers utilize sample that is undisturbed and taken from SRM slope, colluvium, or talus. Due to this variety of the SRM type, methodology, and laboratory sample preparation, a guideline in preparing the sample for laboratory-scale testing is helpful to related practitioners (i.e., geotechnical engineer, engineering geologist) to conduct their experiment and enhance the reliability of the measurement result. Thus, a systematic standard for the soil-rock mixture preparation proposed here shall at least considers and includes the following suggested items (refer to Table 3).

Table 3 Proposed criteria in standard preparation soil-rock mixture

No	Criteria	Description
1	The particle size in the soil-rock mixture.	The sizing includes the minimum and maximum particle size used.
2	Range of acceptable soil-rock ratio that represents the soil-rock mixture.	The soil-rock ratio designed based on the limit on the percentage of soil and rock block.
3	Type of geomaterial used in forming soil-rock mixture.	Soil type based on particle size distribution. Rock block properties include its strength, texture, and shape.
4	Compaction degree.	The percentage of compaction degree must be consistent.

5. Conclusion

Although researchers performed some efforts to assess the strength of SRM based on physical testing in-situ and in laboratory works, further studies to develop more efficient methods and a database for the strength of such materials are still needed. The variation in the soil composition and rock material to prepare the SRM sample allowed us to obtain more information and prepare a comprehensive reference from the various case study. This review conducted successfully enhance the understanding of the variation of sample preparation of SRM and its effect on the strength measurement. The findings in this study also suggest these fundamental factors; moisture content, type of soil, and rock block strength to be sensibly determined in the sample preparation process. The future study proposed considering this input in the development and improvement of a standard systematic procedure of SRM laboratory testing, which is beneficial to the related practitioner.

Acknowledgements
Special thanks to the research supervisor, Assoc. Prof. Dr Ir. Hisham Mohamad for his guidance to finish this paper.

References

[1] Zhang S, Tang H, Zhan H, Lei G and Cheng H 2015 *Int. J. Rock Mech. Min. Sci.***77** 208-219
[2] Xu W J, Hu L M and Gao W 2016 *Int. J. Rock Mech. Min. Sci.* **86** 166–178
[3] Zhao Y and Liu Z 2018 *Adv. Civ. Eng.* 2018 10
[4] Zhang Z, Sheng Q, Fu X, Zhou Y, Huang J and Du Y 2020 *Bull. Eng. Geol. Environ.* **79** 2423–2437
[5] Li Z, Hu F, Qi S, Hu R, Zhou Y and Bai Y 2020 *Appl. Sci.* **18** 6475
[6] Zhang Y, 2017 *Chem. Eng. Trans.* **59** 985–990
[7] Zhang H Y, Xu W J and Yu Y Z 2016 *Soils Found.* **56** 44–56,
[8] Wang S, Ji T, Xue Q, Shen Z and Zhang Q 2019 *Adv. Mater. Sci. Eng.* 2019 **13**
[9] Li B 2020 *E3S Web Conf.* **165** 1–6
[10] Ren M, Zhao G, Qiu X, Xue Q and Chen M 2018 *Adv. Civ. Eng.* 2018 **9**
[11] Xu W J and Zhang H Y 2021 *Bull. Eng. Geol. Environ.* **80** 2715–2726
[12] Li Y 2013 *Bull. Eng. Geol. Environ.* **72** 371–381
[13] Sun S, Zhu F, Wei W, Wang and Le H 2019 *Shock Vib.* 2019 **25**
[14] Cen D, Huang D and Ren F 2017 *Acta Geotech.* **12** 391–413
[15] Zhang Z L, Xu W J, Xia W and Zhang H Y 2016 *Int. J. Rock Mech. Min. Sci.* **86** 317–322
[16] Xue Y, Yue L, and Li S 2015 *J. Eng. Geol.* **23** 21–29
[17] Gao W, Hu R, Oyediran I A, Li Z, and Zhang X 2014 *Geotech. Geol. Eng.* **32** 1329–1338
[18] Shaorui S et al. 2014 *Bull. Eng. Geol. Environ.* **73** 123–140
[19] Yu L and Wang Z Y 2013 *Appl. Mech. Mater.* 423–426 1001–1005
[20] Xu W, Qiang X and Rui-Lin H 2011 *Int. J. Rock Mech. Min. Sci.* **48** 1235–1247
[21] Wickland B E, Wilson G W, Wijewickreme D, and Klein B 2006 *Can. Geotech. J.* **43** 928–945
[22] L E Vallejo 2001 *Can. Geotech. J.* 38 1097–1104
[23] Ding X H, Zhou W, Lu X and Gao Y 2018 *J. Cent. South Univ.* **25** 3085–3097
[24] Vallejo L E, Lobo-Guerrero S and Seminsky L F 2018 *Geotech. Spec. Publ.* **234** 74–83
[25] Amini Y, Hamidi A and Asghari E 2013 Shear Strength Characteristics of an Artificially Cemented Sand-Gravel Mixture King Mongkut's University of Technology Thonburi
[26] Yang J, Cai X., Pang Q, Guo X W, Wu Y L and Zhao J L 2018 *Adv. Mater. Sci. Eng.* 2018
[27] Lv X and Zhou H 2018 *J. Geophys. Eng.* **15** 2221–2234
[28] Wang Y, Li X, and Zheng B 2017 *Soil Dyn. Earthq. Eng.* **93** 7–17
[29] Hamidi A, Yazdanjou V, and Salimi N 2009 *Int. J. Geotech. Eng.* **3** 29–38
[30] Selvam B and Chakravarthy K 2015 *Int. J. Latest Trends Eng. Tech. Stress.* **6** 1–8
[31] Qi C, Li L, Wei J and Liu J 2016 *Adv. Mater. Sci. Eng.* 2016
[32] Huang X, Zhou A, Wang W and Jiang P 2020 *Sustain.* **12** 1–18
[33] Yang H, Zhou Z, Wang X and Zhang Q 2015 *Adv. Mater. Sci. Eng.* 2015
[34] Zhao M H, Liu X P and Peng W X 2007 *Rock soil Mech.* **7** 1323–1327

IOP Conf. Series: Materials Science and Engineering 1153 (2021) 012014 doi:10.1088/1757-899X/1153/1/012014

Optimization of deep soil mixing mass for construction of tunnel cross-passage in Bangkok

Mengly Seng[1] and Tirawat Boonyatee[1]

[1] Department of Civil Engineering, Chulalongkorn University, 254 Phayathai Rd, Wang Mai, Pathum Wan District, Bangkok 10330, Bangkok, Thailand

E-mail: menglyseng@gmail.com

Abstract. A cross-passage from a tunnel to an intervention shaft in Bangkok subsoils was modelled using three-dimensional finite element analysis. The optimization was performed for reducing the volume of soil-cement mass while maintaining ground stability. The result showed that the factor of safety increases linearly with the thickness of improved ground. The relationship between the improved groundmass and ground surface settlement curve was made. It was found that the predicted settlement profile can be captured by the formula proposed by Peck and O'Reilly, providing that the empirical coefficient of 0.6 is fit well with the results from the 3D simulation. This value was obtained from the average of empirical coefficient value 0.4 and 0.7, which are the empirical coefficient values of stiff clay and soft clay, respectively based on the overburden thickness.

1. Introduction

The ground stabilized method has to be used to reduce the large deformation and cross-passage stability when cross-passages were constructed in Bangkok. The deep soil mixing method is an effective method used widely for excavation support, particularly for soft clay deposits. The deep soil mixing method enhances the soft ground into soil-cement, which is durable and can be employed at low cost.

Currently, the improvement scheme of the deep soil mixing is not optimal since the improvement zone was conducted from the surface to reach the cross-passage leading to waste much cement consumption. The construction cost can be reduced further by optimizing the deep soil mixing mass.

In this paper, the optimization of deep soil mixing was conducted to achieve stability with deformation requirement using 3D finite element method. Furthermore, the effect of improvement pattern on the tunnel cross passages is also evaluated.

2. Methodology

2.1. *Pattern arrangement of soil-cement mixing*

There are four types of pattern mentioned: single type, wall type, grid type, and block type [1]. The block type of the pattern was selected for this study (Figure 1) since block type is safer and suitable for permeable prevention since it contains the overlapping of each column in all direction which is acted as water barrier [2]. The diameter of 0.7m with the overlaps of columns is about 0.1m, which is connected according to the research on road tunnel and soil-cement wall in Bangkok.

IOP Conf. Series: Materials Science and Engineering 1153 (2021) 012014 doi:10.1088/1757-899X/1153/1/012014

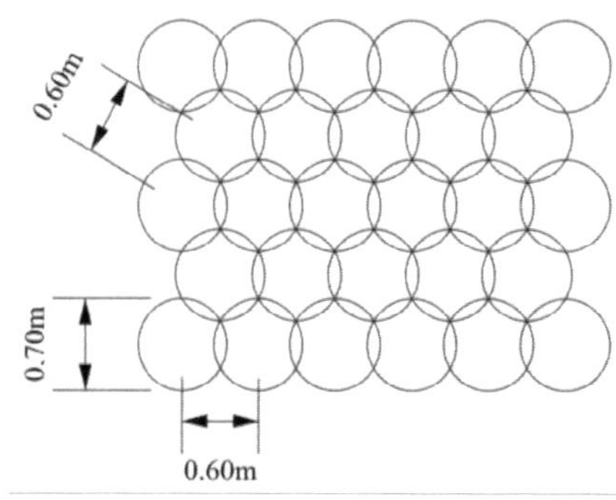

Figure 1. Soil-cement columns in block type pattern

2.2. Optimizing reinforcement

The optimization was based on the stability of the cross-passage where safety factor of 1.40 was considered. The improvement material's optimization is conducted by selecting the optimal improvement radius (R_i) of the ground stabilization, ranging from between 1.2*R* and 3.0*R* while the radius R is 2.25m. The octagonal shape was adopted; however, soil-cement mixing cannot make a perfect oblique pattern (Figure 2).

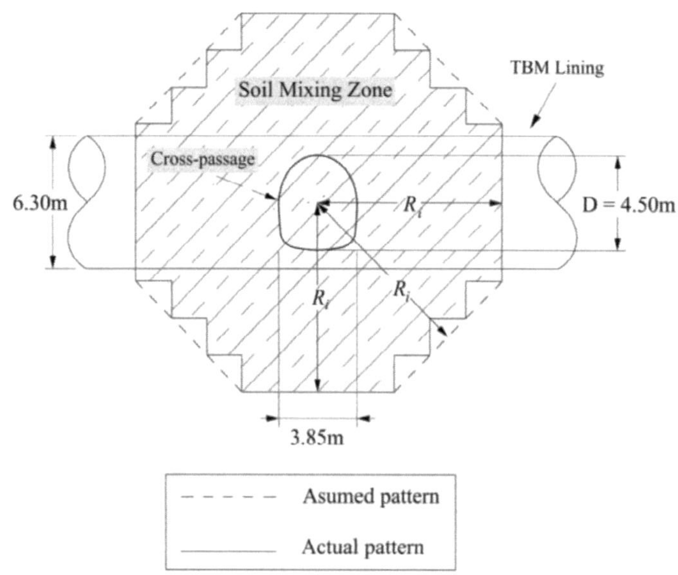

Figure 2. Improvement shape in cross-section view

2.3. Simulation of cross-passage

2.3.1. Subsoil and groundwater condition in Bangkok. The study is based on the subsoil profile in section A of the MRT blue line in Bangkok (Figure 3) by simplified into three layers:

- Layer ⟨1⟩ is denoted as Bangkok soft clay – medium clay layer from the surface to -14m long.
- Layer ⟨2⟩ is denoted as a stiff clay layer from depth -14m to -24m long.
- Layer ⟨3⟩ is denoted as the first sand layer which is from -24m to downward.

Figure 4 indicates the most recent groundwater level in Bangkok subsoil. Since the groundwater in Bangkok is undergone drawdown incredibly, although the current groundwater is increased, it has not fully recovered.

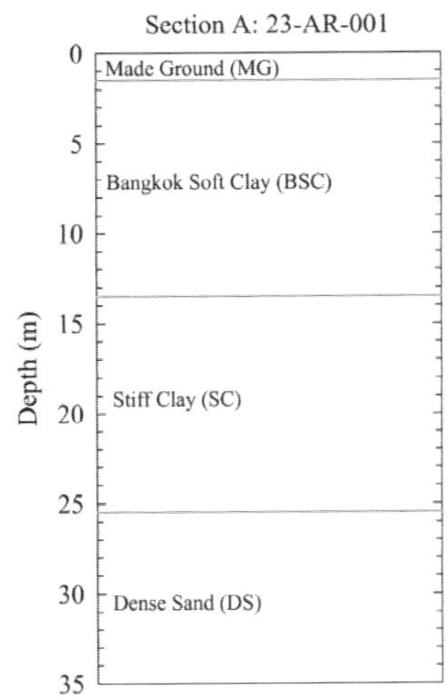

Figure 3. Soil profile in section A of the MRT blue line [3].

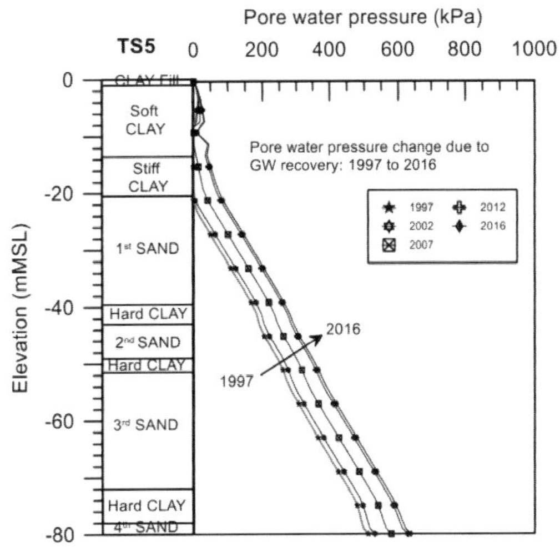

Figure 4. Groundwater recovery [4]

2.3.2. Model geometry and boundary condition. Möller proposed the guideline for 3D modelling in the tunnel for the tunnels diameter range between 4 ~ 12m, the width of the geometry can be suggested in equation (1) [5]. Meißner recommended the bottom boundary is followed by equation (2) [6].

$$w = 2D\left(1 + \frac{H}{D}\right) \tag{1}$$

$$h = (1.5 \sim 2.5) \times D \tag{2}$$

where
w = width of mesh generation
H = distance from the ground surface to tunnel crown
D = tunnel diameter
h = bottom boundary (Starting from the invert of the tunnel down to the considerable length)

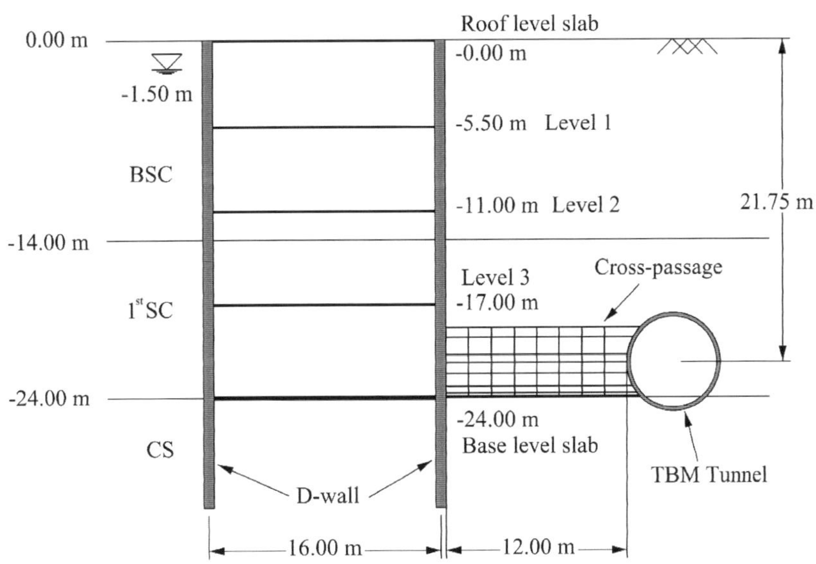

Figure 5. Dimension of the IVS and cross-passage

The layout of the cross-passage, D-wall and TBM tunnel in the model have been illustrated in Figure 5. The cross-passage's width and height are about 3.85m and 4.5m respectively, with 12m long. Cross-passage was modelled at -24m from the surface to bottom of cross-passage while -21.75m from the surface to the centre of the cross-passage and it was connected between TBM tunnel and IVS (16m × 16m).

2.3.3. Soil constitutive and input parameters. Hardening soil model (HSM) has been selected for this research, and the input parameters are adopted from [7], as shown in Table 1.

Table 1. Hardening soil parameters [7].

Soil type	γ_b (kN/m^3)	c' (kPa)	ϕ' (°)	ψ' (°)	E_{50}^{ref} (MPa)	E_{oed}^{ref} (MPa)	E_{ur}^{ref} (MPa)	v_{ur}	m	K	Rf	Analysis Type
BSC	16.5	1	23	0	0.8	0.85	8.0	0.2	1	0.7	0.9	Undrained (A)
1st SC	19.5	25	26	0	8.5	9.0	30	0.2	1	0.5	0.9	Undrained (A)
CS	19.0	1	27	0	38	38	115	0.2	0.5	0.55	0.9	Drained

Mohr-coulomb soil model has selected for the soil-cement material model, and the input parameters are adopted from [8] as shown in Table 2.

Table 2. Soil-cement parameters [8].

Material Type	Analysis Type	v	γ_b (kN/m3)	C (kPa)	E (MPa)
Soil-cement	Undrained (B)	0.33	15	400	0.8

Structures were modelled in plate elements, and structural parameters which are used in this model are adopted from [9] [10] (Table 3).

Table 3. Parameters of the tunnel lining and intervention shaft [9] [10].

Parameters	d (m)	γ (kN/m^3)	$E1$ (GPa)	v
Tunnel lining	0.30	24	31	0.20
D-wall	1.00	16.5	28	0.15
Slab	1.00	25	28	0.15
Base slab	1.80	25	28	0.15

2.3.4. Construction sequences. The simulation was categorized into six main stages — first, the initial phase where K_0 is selected (a). Secondly, soil-cement mixing columns were constructed around the excavation area (b). Then the TBM tunnel was activated (c). Next, the intervention shaft was activated (d). After then the excavation was operated every -1.5m to final excavation (e) (f). Full face excavation is conducted in this study, and face pressure is not applied to the tunnel (Figure 6).

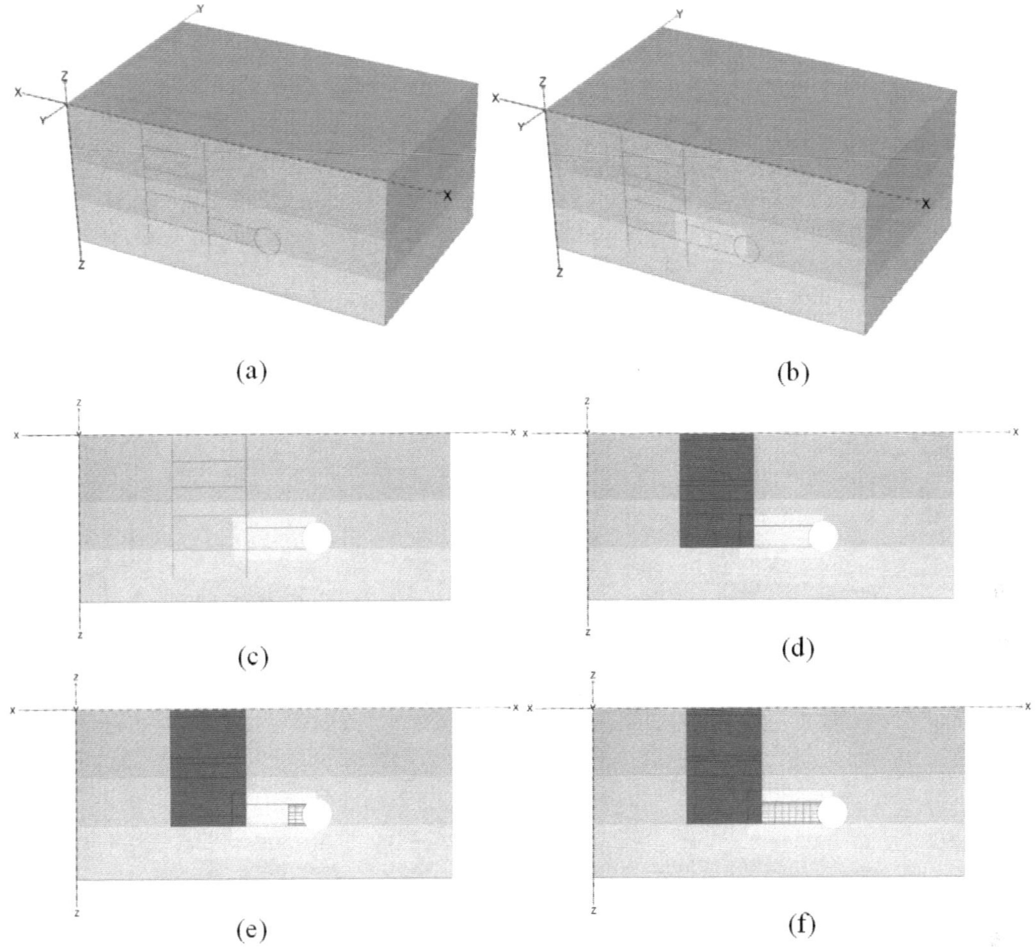

Figure 6. The process of cross-passage in PLAXIS 3D: (a) Initial Phase, (b) Soil-cement mixing, (c) TBM tunnel was constructed, (d) IVS was constructed, (e) Cross-passage open face and excavation, and (f) Final excavation.

2.4. Model Validation

2.4.1. Ground movements induced by tunnelling. An immediate ground surface settlement profile above a single tunnel can be represented with the empirical equation below [11] [12]:

$$\delta = \delta_{\max} \cdot \exp\left(-\frac{x^2}{2i^2}\right) \tag{3}$$

where
δ = surface settlement at a lateral distance
δ_{max} = maximum surface settlement at the centerline
x = distance from the tunnel centerline

The settlement trough equation was updated which is respective to the depth of the tunnel and the type of the ground as following in equation (4). O'Reilly proposed that the empirical coefficient (K) is selected based on the types of soil overlaid on the tunnel [13].

$$i = KZ_o \tag{4}$$

where
i = distance from the tunnel centerline to inflection point
K = empirical coefficient
Z_o = depth to the tunnel springline

2.4.2. Effect of ground improvement on tunnel stability. The ground improvement design was conducted in the plastic region of the soil where the soil was enhanced by increasing the shear strength [14]. Figure 7 (a) shows that when the shear strength was increased, there is no ground failure unless the achieved shear strength touches a failure line of the Mohr circle of the original ground. The radial and tangential stress are equal to each other at the elastic region's boundary from the plastic region, as shown in Figure 7 (b).

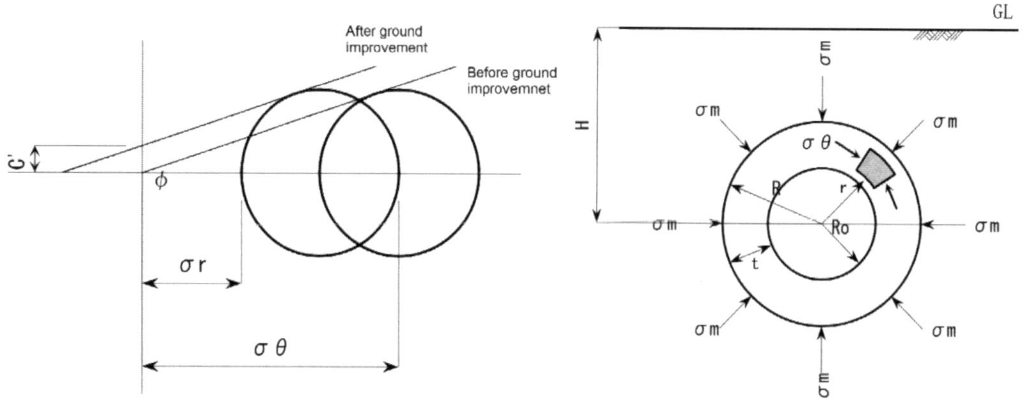

(a) (b)

Figure 7. (a) Mohr circle of the improved ground and original ground; (b) Cross-section of improvement zone based on the plastic theory [14]

The equation of the radial plastic region can be obtained lead to the following:

$$\ln R + \frac{R \cdot \gamma_t}{2C} = \frac{H \cdot \gamma_t}{2C} + \ln R_0 \tag{5}$$

$$t = FS \cdot (R - R_0) \tag{6}$$

where
R = radius of plastic region
R_0 = radius of tunnel
γ_t = average unit weight of soil
C = shear strength of improved soil
H = depth to the centre of the tunnel
FS = factor of safety
t = thickness of the improvement zone

3. Results and discussion

3.1. Stability of cross-passage

Figure 8 shows the results of the safety factor, which increase linearly with the improvement thickness. The round marks are denoted as the PLAXIS 3D model results while the triangle marks are denoted as the improvement thickness of improvement based on the plastic theory [14]. The analytical formula of Shibazaki using for improvement in the TBM tunnel with circle improvement zone while the pattern of the soil improvement in this research is octagonal. From the graph, the safety factor of the 1.6R case is 1.70, which is acceptable since the safety factor is more than 1.40.

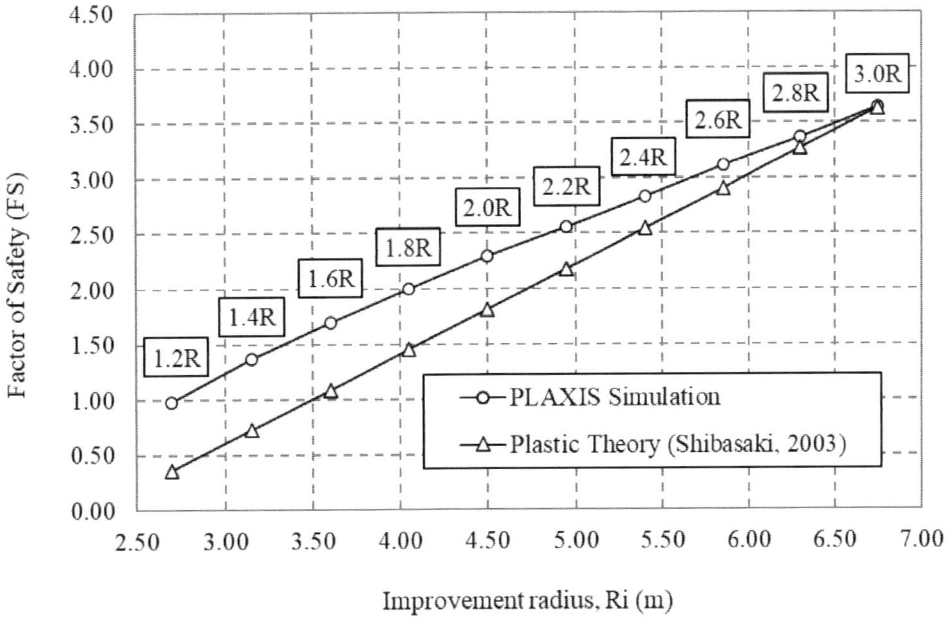

Figure 8. Stability of the cross-passage

3.2. *Behaviour of ground surface displacement*

The result of the settlement profile from FEM is agreed well with the empirical equation of O'Reilly and New [13]. In the figure, three types of line represented the empirical coefficient varying depending on the soil types. The K value ranges from 0.4 (stiff clay) to 0.7 (soft clay) for the cohesive soil. In this study, $K = 0.6$ fits well with the results of 3D simulation where $K = 0.6$ is obtained from the average of K value based on the thickness of each type of soil above the crown of cross-passage. Overall, the empirical coefficient $K = 0.6$ is the suitable value for this case, and it also can be proved that the displacement from the 3D simulation is accurate.

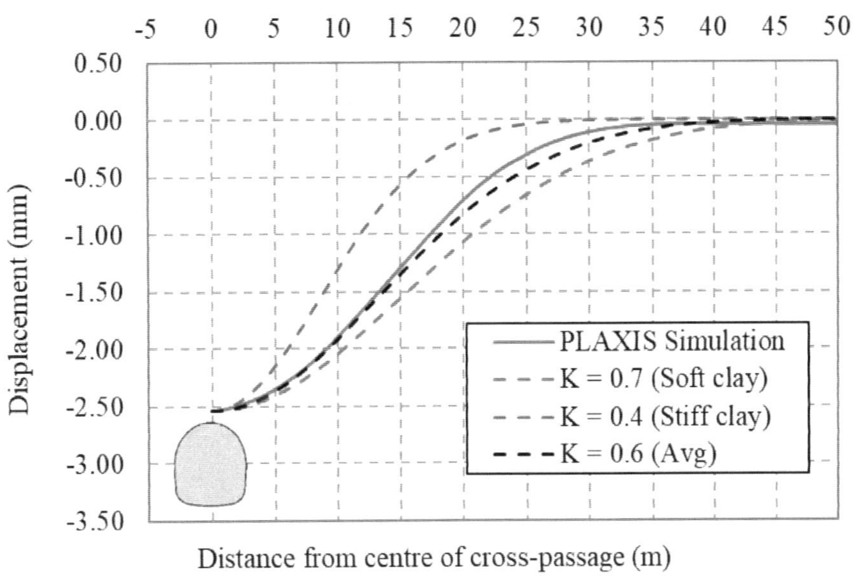

Figure 9. Ground surface profile induced by tunnel excavation

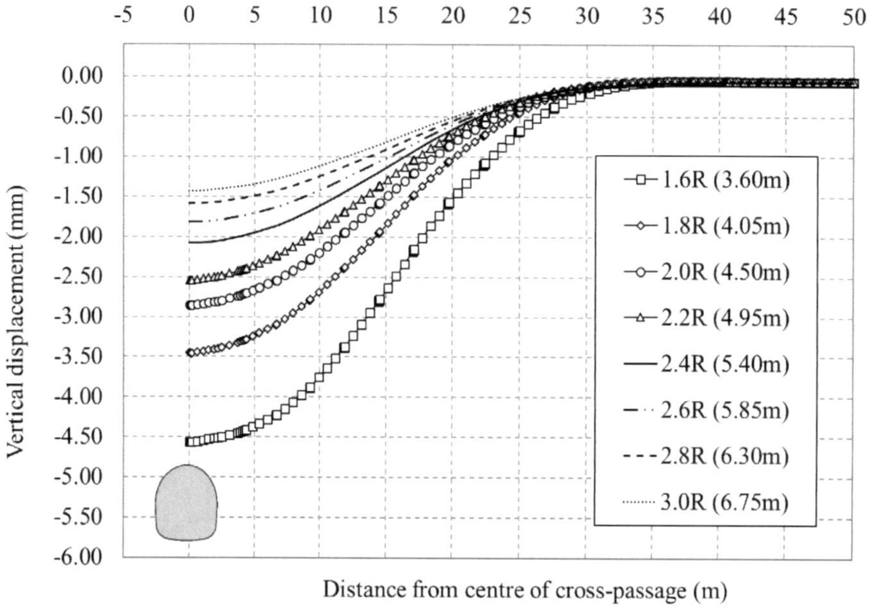

Figure 10. Comparing the surface settlement of improvement schemes 1.6R to 3.0R

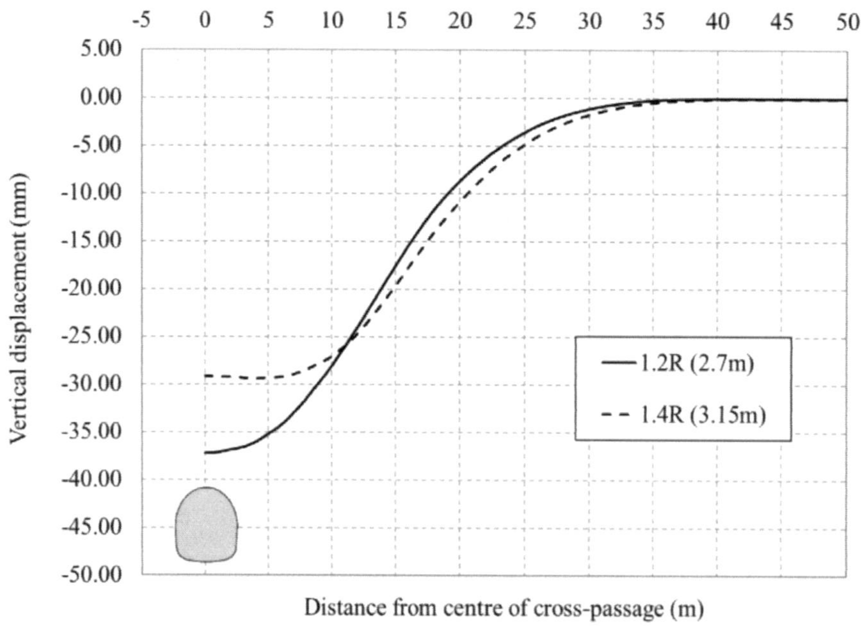

Figure 11. Comparing the surface settlement of improvement 1.2R and 1.4R.

Figure 10 indicates the ground surface behaviour after the soil being improved with the improvement thickness increased from 1.6D to 3.0D. The settlement of the 1.6D improvement scheme is about 4.5mm while the settlement is decreasing significantly to less than 1.5mm in the 3.0D scheme. However, the improvement scheme of 1.2R and 1.4R shows that the vertical displacement is extensive, about ten times those, see Figure 11. The improvement scheme of 1.2R found that soil collapsed at 4.5m exaction.

4. Conclusion

Optimization based on the safety factor indicated that the ground improvement 1.6R scheme was the best choice because the safety factor is more than 1.40.

The ground surface displacement of the 1.6R improvement scheme was about 4.57mm. The ground surface settlement can be reduced by increasing the improvement radius. The validation of ground surface displacement of the finite element model and the empirical equation of O'Reilly was found to fit well when the empirical coefficient $K = 0.6$ was used.

References

[1] Bruce D A 2000 An introduction to the deep soil mixing methods as used in geotechnical applications Turner-Fairbank Highway Research Center 6300 Georgetown Pike McLean, VA 22101
[2] Kitazume M and Terashi M 2013 The deep mixing method: CRC press
[3] Likitlersuang S, Surarak C, Suwansawat S, Wanatowski D, Oh E and Balasubramaniam A 2014 Simplified finite-element modelling for tunnelling-induced settlements. *Geotechnical Research* **1** pp 133–152
[4] Saowiang K and Giao P H 2020 Numerical analysis of subsurface deformation induced by groundwater level changes in the Bangkok aquifer system. *Acta Geotechnica*
[5] Möller S C 2006 Tunnel induced settlements and structural forces in linings. The University of Stuttgart, Institute of Geotechnical Engineering, Stuttgart, Germany
[6] Meißner H 1996 Tunnelbau unter Tage. *Empfehlungen des Arbeitskreises* **1** p 99
[7] Likitlersuang S, Surarak C, Wanatowski D, Oh E and Balasubramaniam A 2013 Finite element analysis of a deep excavation: A case study from the Bangkok MRT. *Soils and Foundations* **53** pp 756–773
[8] Jamsawang P, Yoobanpot N, Thanasisathit N, Voottipruex P and Jongpradist P 2016 Three-dimensional numerical analysis of a DCM column-supported highway embankment. *Computers and Geotechnics* **72** pp 42–56
[9] Chheng C and Likitlersuang S 2018 Underground excavation behaviour in Bangkok using three-dimensional finite element method. *Computers and Geotechnics* **95** pp 68–81
[10] Lueprasert P, Jongpradist P, Charoenpak K, Chaipanna P and Suwansawat S 2015 Three dimensional finite element analysis for preliminary establishment of tunnel influence zone subject to pile loading. *Maejo International Journal of Science and Technology* **9** pp 209–223
[11] Peck R B 1969 Deep excavations and tunnelling in soft ground. *Proceedings of the Seventh International Conference on Soil Mechanics and Foundation Engineering*, Mexico, State of the Art Report pp 22–290
[12] Schmidt B 1969 Settlements and ground movements. Department of Civil Engineering, University of Illinois, Urbana, Illinois, 224
[13] O'Reilly M and New B 1982 Settlements above tunnels in the United Kingdom-their magnitude and prediction. *Proceedings of the Third International Symposium, Tunnelling* pp 173–181
[14] Shibazaki M 2003 State of practice of jet grouting *Grouting and Ground Treatment* pp 198–217

Acknowledgements

My deepest thanks to my advisor, Assoc. Prof. D.Eng. Tirawat Boonyatee for his guidance and instruction of this research and provide me with so much knowledge. I want to extend my heartfelt appreciation toward lecturers and researchers in Geotechnical Engineering Division Chulalongkorn University for their valuable knowledge and instruction during my study. Furthermore, I would like to show my sincere thanks to my parents, friends and seniors who help, support, and give me good advice. Lastly, I am also thankful to Chulalongkorn University ASEAN scholarship, which supports the fund for my research and study.

Effects of Cement on the Compaction Properties of Lateritic Soil

N A Wahab[1], A S A Rashid[1], M J Roshan[1], N H A Rizal[1], N Z M Yunus[1], M A Hezmi[1], M Y M Tadza[2]
[1]School of Civil Engineering, Universiti Teknologi Malaysia Skudai, Malaysia
[2]Faculty of Civil Engineering, Universiti Malaysia Pahang, Malaysia
Email: norshakila@graduate.utm.my

Abstract. Lateritic soils at Universiti Teknologi Malaysia (UTM) Skudai, Malaysia was investigated subjected to their geotechnical properties and their suitability to be used as construction materials. The soil samples were collected and tested for various basic soil properties tests such as Atterberg Limit, Specific Gravity (SG), Sieve Analysis, and Compaction test. Atterberg's consistency limit test shows that the liquid limit (LL) is 70.3%, the plastic limit (PL) is 42.0%, and the plasticity index (PI) is 28.3%. The specific gravity (SG) value for laterite is 2.74. For the compaction test, the optimum moisture content (OMC) and maximum dry density (MDD) obtained are 28% and 1.39 g/cm^3. A laboratory study was performed to compare and evaluate the stabilization efficiency of different percentages (3%, 6%, 9%, 12%) of Ordinary Portland Cement (OPC) when applied to the available laterite soil; a major soil group in the tropical areas. Analysis of laboratory data is assessed from a soil compaction test through the standard proctor method by using the automatic compactor. The soil mixtures were compacted at optimum moisture content in accordance with the British Standard (BS) of BS 1377-4:1990. From the preceding results, it was found that the OMC increase from 28% to 34% while the MDD increase from 1.39 g/cm^3 to 1.47 g/cm^3 with the rise in the percentage of cement.

1. Introduction

It has been estimated that road lengths in Malaysia are around 216,837 kilometres which are comprised of federal and state roads [1]. Mainly, the roads are categorized under low volume road type enclosed by rural development as it promotes access to people for generating their income. But due to the rapid population growth, Malaysia is facing an insufficiency of desirable road infrastructure. Therefore, the soil stabilization method may be used in the process of low-volume road enhancement in Malaysia to achieve a self-sustaining nation.

Lateritic soils are soil groups rich in iron where they are formed in hot and wet tropical regions. Generally, all lateritic soils are developing by intensive and severe weathering of the parent rock, and mostly they are in reddish-brown because of the high iron oxides content. Laterite soils comprise about one-third of the earth's continental land sections [2]. Besides, lateritic soils are shaped from the leaching of parent sedimentary rocks (clays, limestones, sandstones), volcanic rocks, and mineralized rocks, which permit the more enigmatic irons, mostly iron and aluminium [3].

Normally, laterite soils have been used as subgrade layers for low-volume road pavement because they are easy to operate on the road surface and naturally have grading stability [4]. Besides, they also have a certain proportion of clayey materials, which act as binders in pavement layers. Therefore, to attain a sustainable road, the process of soil stabilization shall be implemented as stated by various researchers such as [5], [6], [7], and [8].

Soil stabilization is extensively known as soil treatment which allows the soils in advancing their strength and durability such that they will be beyond their original properties in construction purposes. Soil stabilization techniques may include the blending or mixing of soils with the available additives or agents that may alter the texture, gradation, or plasticity of soil [6]. Soil stabilization can be conducted through various techniques. All these techniques fall into two major groups, which are chemical and mechanical stabilization. Chemical stabilization is an alteration of soil geotechnical properties through a chemical reaction between chemical additives (lime and cement) and soil minerals (pozzolanic material) [9]. Mechanical stabilization is not the focus of this study; therefore, it will not be further discussed.

Cementitious stabilization is a type of chemical stabilization, which has been used widely and broadly in Malaysia. For instance, cement-stabilized soil was used as a road base instead of mine gravel at Labuk and North Roads in Sabah [10]. A model of cement-stabilized soil also was constructed by [11] to study the effect of weathering, soil-cement had been applied instead of using crushed aggregate in a road development at Labuan and Sandakan, Sabah [12], and cement stabilization of high fines in Melaka can achieve the compressive strength of 2.8 MPa through the fines content modification with river sand [13].

Soils are deemed as unique materials which they do not always show the desired properties for constructions. Hence, soil modification is necessary at the site, to enhance their engineering properties. Due to its cost-effectiveness, soil compaction is the widely used method to achieve soil properties enhancement. By compacting the earth fills, the subsequent soil settlement may be reduced, the soil permeability may be decreased, and the shear strength of the soil also can increase. Compaction is essential in many geotechnical applications such as railway subgrades, road pavements, and earth retaining structures. However, laboratory compacted soil is mostly used for pavement design.

Therefore, the main objective of this research is to compare the compaction characteristics of untreated and cement-treated lateritic soil. Besides, the author also intends to investigate the impact of cement percentage on the compaction behavior of soil.

2. Materials and Method

For the study, the disturbed samples were collected from a borrow pit nearest to Block P16, Faculty of Electrical Engineering at Universiti Teknologi Malaysia Skudai, where 40 kilograms of laterite soils were excavated using a backhoe, quickly placed in an airtight container, and brought to the laboratory for further laboratory analysis which includes: Natural Moisture Content, Atterberg Limit, SG, Sieve Analysis and Compaction test. Lateritic soils at Universiti Teknologi Malaysia (UTM) Skudai, Malaysia was investigated subjected to their geotechnical properties.

While cement that is used in this research is from the Holcim brand of OPC (CEM I 42.5N) which is manufactured by YTL Cement Sdn. Bhd. From the literature, it is deemed that cement acts to be a useful additive material. Therefore, this laboratory method is based on the formulation of the mixture between different cement percentages with poor material (lateritic soil) to produce a better quality product. The constituents of the OPC are described in **Table 1**, which took from the provided cement test certificate by YTL Cement Sdn. Bhd.

Table 1. Cement specifications from the test certificate of the Holcim OPC.

Specification Items	MS EN 197-1:2014 (CEM I 42.5N) SPECIFICATIONS	Test Results
Initial setting time (minutes)	MS EN 196-3:2007 Clause 6	140
Soundness : Expansion (mm)	MS EN 196-3:2007 Clause 7	1
Compressive strength (N/mm^2) – 2 days	MS EN 196-1:2007	25.8
Compressive strength (N/mm^2) – 28 days	MS EN 196-1:2007	52.4
Loss on ignition, LOI (%)	MS EN 196-2:2007 Clause 7	0.97
Insoluble residue, IR (%)	MS EN 196-2:2007 Clause 9	1.24
Chloride, Cl (%)	MS EN 196-2:2007 Clause 14	0.03
Sulphate, SO_3 (%)	In house test method SOP-QCD-PGP-066	3.09

All the laboratory testing procedure starts with three days of air-dried process of the soil samples at the provided place, with a temperature of 26°C. After that, the procedure continues by drying the samples in the oven for three days at 60°C. The reason for this low-temperature drying technique is to avoid the desiccations of organic material and oxidation of the mineral elements in the lateritic soils, as well as mentioned by [14] in their research.

The grain size distribution is performed by a combination of wet sieving and particle analyzer method according to the [15] as well as the Atterberg limit and SG tests. While for the test that comprising moisture-density relationship is conducted on both untreated and treated lateritic soil according to the [16]. In the compaction test, the cement addition (3%, 6%, 9%, and 12% by dry weight of soil) was blended into the soil-water mixture with thoroughly mixing until a homogeneous wet cemented soil mixture was achieved.

After that, the soil is compacted into a 1000 cm³ volume of proctor mould by using an automatic compactor machine. The compaction effort performed throughout the laboratory test is the modified proctor method where the energy derived from 4.5 kg rammer falling through onto three layers which each is receiving 25 blows. Next, the samples are extracted by using the modified extruder. However, it is essential to note that, due to the utilization of the automatic mechanical compactor, the result dispersions do not subject to the compaction energy applied.

$$\rho_b = W/V \tag{1}$$

Where ρ_b is the bulk density, W is the weight of compacted soil and V is the volume of the mould. After that, from Equation 1, dry density is obtained.

$$\rho_d = \frac{\rho_d}{1+(\frac{MC}{100})} \tag{2}$$

Where ρ_d is the dry density and MC is the moisture content.

By compacting the soil through the modified proctor method, subsequent settlement may be reduced under working loads as well as decreasing the void gaps between soil particles because it creates difficulties for water to freely move and flow through the compacted soil.

3. Results and Interpretations

The engineering properties of lateritic soil have been obtained, as shown in **Table 2.** From the obtained particle size results, the soil tends to be clay and also considered as high plastic because the

PI value is more than 17 [17]. Besides, it shows that the specimen is sandy elastic silt following the USCS (Unified Soil Classification System) based on the grading curve results. The SG values of soil are quite high, which indicating that the soil is tropical iron-rich laterite. Compaction results for untreated soil are shown in **Figure 1,** where the OMC is 28%, and MDD is 1.39 g/cm³.

Figure 2 displays the compaction curves of laterite and cemented laterite specimens with 3%, 6%, 9%, and 12% of cement addition. The cemented soil exhibited greater values of MDD than the untreated one while the OMC fluctuates a little (**Figure 3**). This expresses a universal indication of soil enhancement. For instance, the MDD is increased from 1.39 g/cm³ (natural laterite) to 1.41 g/cm³ (3% soil-cement) and keep increased until 1.47 g/cm³ (12% soil-cement).

It is deduced that the significant increment in MDD is because of the cement addition since the cement tends to loosen the soil particles in the presence of water, which causing denser packing during the compaction process. Hence, the increment of MDD in soil-cement mixtures is due to cement particles that performed cations exchange with the soil elements, thus occupying the voids and compactly packing the soil particles together. Besides, the fine cement particles tend to fill up the void spaces between the soil particles, which developing a denser soil matrix.

While for the OMC, it is increased from 28% to 30%, then it reduced to 28% and increased again until 33%. A similar trend has also been reported by [18], [19], [20], and [21] in their research. These patterns explained that an increase in OMC of soil-cement mixtures is because of the cementitious binder increment in the soil, facilitating the movement of ions in the pozzolanic reaction. However, the slight decrease in OMC at 9% soil-cement (from 30% to 28%) may be due to the self-degradation of the water used during the cement hydration process.

The zero air void line (ZAVL) present in the graph (**Figure 2),** acts as the baseline for the compaction results. ZAVL is well-known as a line which plots from the dry unit weight of soil versus moisture content at 100% degree of saturation. Therefore, if ZAVL overlaps the compaction line, that probably indicates the unreliability results; this is because, during the proctor test, air voids cannot be removed completely. Hence, the compaction curve can't cross the ZAVL as ZAVL is the plotted line without any air voids in the soil.

Table 2. Properties of lateritic soil.

Natural Moisture Content (%)	40.34
Liquid Limit (%)	70.3
Plastic Limit (%)	42.0
Plasticity Index (%)	28.3
Gravel (%)	13
Sand (%)	18
Silt and Clay (%)	69
C_u	8.96 (well-graded)
C_c	0.98 (well-graded)
USCS	MH (Sandy Elastic Silt)
Specific Gravity	2.74
OMC (%)	28
MDD (g/cm³)	1.39

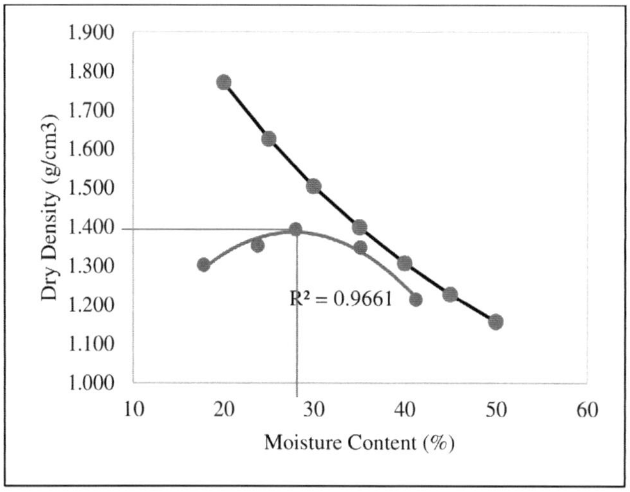

Figure 1. MDD vs OMC graph for untreated lateritic soil.

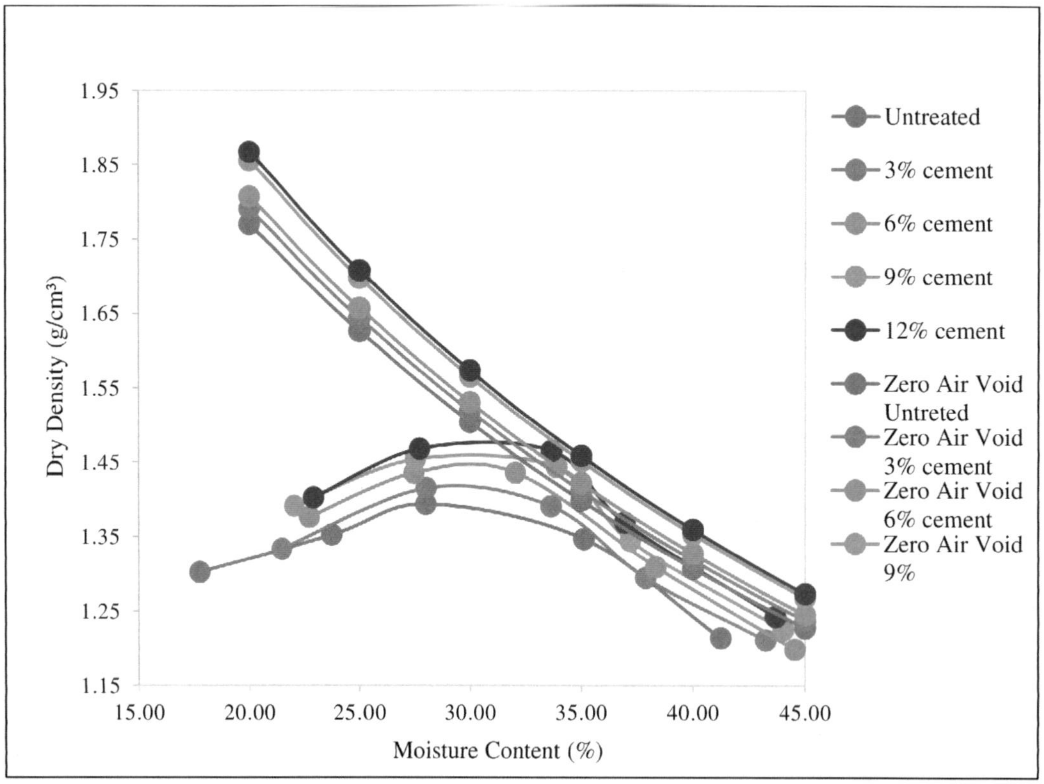

Figure 2. MDD vs OMC graph for untreated and treated lateritic soil.

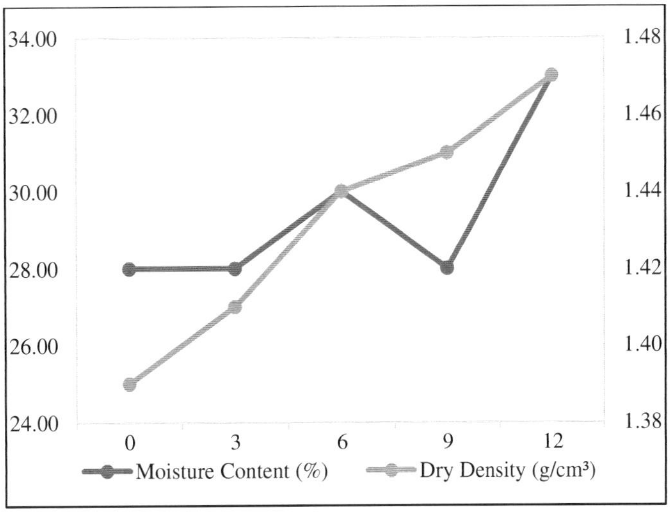

Figure 3. Variations of DD and MC values for untreated and treated lateritic soil.

4. Conclusion

This paper investigated the influence of cement on the compaction characteristics of lateritic soil (OMC and MDD). Based on the obtained results, the following conclusions can be drawn:

1. The maximum dry density (MDD) increased with cement content, indicating that the addition of cement causes the specimen to denser.
2. The optimum moisture content (OMC) increased with cement content, implying that the more cement added, the more water is necessary.
3. The zero air void line (ZAVL) did not cross the compaction curves, indicating the correctness and reliability of the compaction results.

Acknowledgement

This work was sponsored by the Fundamental Research Grant Scheme (FRGS) awarded by the Ministry of Education of Malaysia (Engineering and Microstructural Characteristics of Lateritic Soil Treated with Ordinary Portland Cement Under Cyclic Saturated (Wetting) and Unsaturated (Drying) Conditions - R.J130000.7851.5F131).

References

[1] Buku Statistik Jalan (2016). Cawangan Senggara Fasiliti Jalan 2016, Kuala Lumpur: Jabatan Kerja Raya.
[2] Bureau of Reclamation. (1998). Earth Manual, 3rd Edition, Technical Science Center, Colorado, USA, 21-67.
[3] Smith, G. a. (1998). Elements of Soil Mechanics, 7th Edition. Chapter 6.
[4] Rosli, R. N., Selamat, M. R., & Ramli, M. H. (2019). Shear Strength And Permeability Properties Of Lateritic Soils From North West Malaysia Due To Extended Compaction. Materials Today: Proceedings, 17, 630–639.
[5] Saing, Z., Samang, L., Harianto, T., & Patanduk, J. (2017). Study On Characteristic Of Laterite Soil With Lime Stabilization As A Road Foundation. International Journal of Applied Engineering Research, 12(14), 4687–4693.
[6] Makusa, G. P. (2012). Soil Stabilization Methods and Materials in Engineering Practice. Journal, 1, 1–35.

[7] Oluwatuyi, O. E., Adeola, B. O., Alhassan, E. A., Nnochiri, E. S., Modupe, A. E., Elemile, O. O., Obayanju, T., & Akerele, G. (2018). Ameliorating Effect Of Milled Eggshell On Cement Stabilized Lateritic Soil For Highway Construction. Case Studies in Construction Materials.
[8] Sani, J. E., Yohanna, P., & Chukwujama, I. A. (2020). Effect Of Rice Husk Ash Admixed With Treated Sisal Fibre On Properties Of Lateritic Soil As A Road Construction Material. Journal of King Saud University - Engineering Sciences, 32(1), 11–18.
[9] James, J., & Pandian, P. K. (2016). Industrial Wastes as Auxiliary Additives to Cement/Lime Stabilization of Soils. Advances in Civil Engineering, 2016.
[10] AW, S. (1988). Geotechnical And Environmental Associates Sdn Bhd, Malaysia. Personal communication.
[11] Megat Mohd Nor, A. A. (1988). Soil-Cement For Low Cost Roads. Proceedings of the 14th WEDC Conference KL, 103-106.
[12] Loh. (1986). Preliminary Studies Of Soil-Cement As A Road Building Material. Universiti Pertanian Malaysia.
[13] Megat Mohd Noor, M., Abdul Aziz, A., & Maail, S. (1995). Cement-stabilized Modified High Fines Melaka Series for Roadbases. Pertanika Journal of Science & Technology, 3(2), 285–295.
[14] Mengue, E., Mroueh, H., Lancelot, L., & Medjo Eko, R. (2017). Physicochemical And Consolidation Properties Of Compacted Lateritic Soil Treated With Cement. Soils and Foundations, 57(1), 60–79.
[15] BSI. (2014). BSI Standards Publication: Geotechnical investigation and testing — Laboratory testing of soil Part 4 : Determination of particle size distribution. BSI.
[16] BSI. (2014). BSI Standards Publication: Geotechnical investigation and testing — Laboratory testing of soil Part 2 : Determination of bulk density. BSI.
[17] Sowers, G. F. (1979). Introductory Soil Mechanics & Foundations: Geotechnic Engineering.
[18] Mengue, E., Mroueh, H., Lancelot, L., & Medjo Eko, R. (2018). Design And Parametric Study Of A Pavement Foundation Layer Made Of Cement-Treated Fine-Grained Lateritic Soil. Soils and Foundations, 58(3), 666–677.
[19] Marathe, S., Kumar, A., & Avinash. (2015). Stabilization of Lateritic Soil Subgrade Using Cement, Coconut Coir and Aggregates. International Journal of Innovative Research in Science, Engineering and Technology, 4(12), 11907–11914.
[20] Bayat, M., Asgari, M. R., & Mousivand, M. (2013). Effects Of Cement And Lime Treatment On Geotechnical Properties Of A Low Plasticity Clay. International Conference on Civil Engineering Architecture & Urban Sustainable Development 27&28 November 2013, Tabriz, Iran Effects, June.
[21] Xu, F., Wei, H., Qian, W., Chen, X., Xu, T., He, Y., & Wen, G. (2020). Experimental Investigation On Replacing Cement By Sintered Limestone Ash From The Steelmaking Industry For Cement-Stabilized Soil: Engineering Performances And Micro-Scale Analysis. Construction and Building Materials, 235.

IOP Conf. Series: Materials Science and Engineering 1153 (2021) 012016 doi:10.1088/1757-899X/1153/1/012016

Calibration and Validation of Lane Changing Following Gap Distance in VISSIM

H A Mohamed[1], O C Puan[2] , S A Hassan[1] and S F Azhari [3]
[1] School of Civil Engineering, Faculty of Engineering, Universiti Teknologi Malaysia, Johor Bahru, Malaysia
[2] Department of Civil Engineering, Engineering College, Universiti Malaysia Pahang, Malaysia
[3] Building and Road Research Institute, University of Khartoum, Khartoum, Sudan
Email: a.ali1976@graduate.utm.my

Abstract: Lane change (LC) is an essential feature of expressway driving behavior that affects traffic efficiency and safety. Thus, a wrongful execution of LC, could be due to the wrong choice of gap distance before the LC manoeuvre. This might lead to road crashes, mostly in form rear-end, sideswipe, or angled collision. This research evaluates the characteristics of drivers' Following Gap Distance (FGD) during LC using VISSIM traffic simulation program. Data were collected along a representative segment along with Kuala Lumpur – Seremban expressway for two days using an instrumented vehicle approach. A passenger car was instrumented with a Video Velocity Box (VBox), an on-board data acquisition system used for video recording of the road's traffic event. A total of 75 and 20 following incidences of lane changing were observed in first and second day, respectively. Standstill Distance (CC0), Gap Time (CC1) and Following Distance Oscillation (CC2) were used as calibration parameters in VISSIM model. FGD was used as a measure of performance for the model. The data from second day was used for model validation. The findings indicate that the model was well calibrated and validated. Therefore, it can be used for modelling various scenarios involving LC on the expressway.

1.0 Introduction

Lane changing (LC) is referred to as the driving manoeuvre in which a vehicle moves from one lane to another, where both lanes have the same travel direction. An LC is one of the most sensible driving behavior characteristics, which has mostly influences traffic efficiency and safety[1]. Changing lanes happens for various reasons, such as entering the road, planning to leave the road, expecting other vehicles to merge on the highway, anticipating a slower leading vehicle, changing the number of lanes open, and so on.[2]. Wei [3] demonstrated that when a leading vehicle's speed in the current lane is lower than that of its follower, acceptable gaps are available in the target lane. The following driver tends to change lane.

The most common type of lane changing is a maneuver in which a driver changes lanes to pass a slower lead vehicle in order to continue with the current speed[2]. LC performance choice may be affected by some factors, including the emergence of the following vehicle, surrounding traffic impact, and driver's character[4]. A lane change's crash occurs when drivers attempt to change lanes and strikes or struck by other vehicles in the adjacent lane in most incidences. Following Gap Distance FGD is defined as a space interval that a subject vehicle requires to merge safely between two vehicles. On the other hand, FGD is the distance from front bumper of the following vehicle to the leading vehicle's rear bumper. Figure 1 depicts the following gap distance between the test car and the immediate following vehicle. However, an erroneous estimate (by the following driver) of FGD to the leading vehicle and unevaluated adequate time to pass safely are mostly the key reasons for LC crashes.

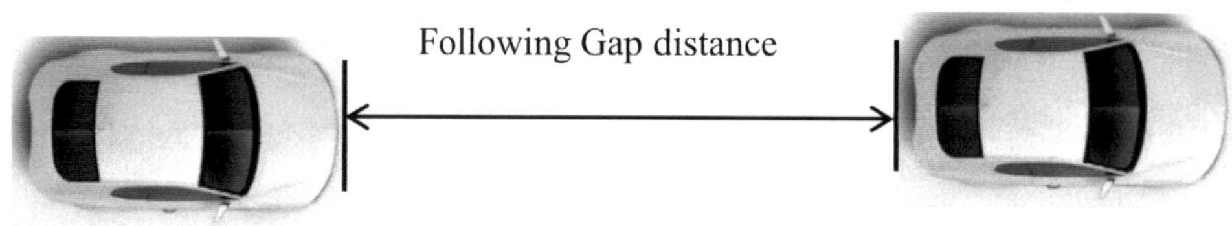

Figure 1. Following gap distance between vehicles

Microscopic traffic simulation has been the common approach used by transportation engineers to conduct traffic impact studies. In traffic simulation models, drivers of typical vehicles' stochastic behavior are delineated using chance distributions to obtain the parameter values[5]. A computer-based approach is used during a simulation phase to evaluate a model numerically, and data is generated to estimate the defined actual mode characteristics[6]. Researchers Srikanth [5], and Mehar [7] have recently shown an increased interest in traffic simulation programs and simulation of lane changing behavior with more emphasis and attention to safety gap distance. More recently, concerns have been raised by some relevant researchers[5], Farooq [8], who developed models via a traffic simulation approach for the determination of the safety of FGD while drivers change lanes.

To date, macroscopic-level lane-changing studies have been mostly centred on two aspects; (a) the relation between the number of lane changes and different traffic state variables, such as traffic flow, mean speed, density, and so forth, and (b) lateral and longitudinal distributions of lane changing. However, microscopic traffic models describe longitudinal car-following and lateral lane-changing behavior of individual vehicles[2].

It is most important to build individual driver behavior models based on the above arguments, which could simulate drivers' behavior, particularly for maneuvers that are susceptible to accidents. This work focussed on the calibration and validation of FGD via the application of VISSIM for motorway or expressway conditions. A behaviour simulation model would be developed that could simulate the findings following the objective somewhat similar to that was examined by Weber[9], and Xiaorui [10] established an LC scenario and derived a reasonable LC model, and subsequently affirmed the model's performance using VISSIM simulation. The results indicated that the FGD is appropriate for the safety LC manoeuvre, which may fit the perfect LC situation.

2.0 Literature Review

Based on homogeneous traffic conditions, various traffic simulation programs have been developed in different countries. VISSIM is one such microscopic traffic simulation model. This simulation software package by *PTV Planung Transport Verkehr AG* in Germany was based on Wiedemann's continuous work. VISSIM is a microscopic, time step and behaviour based simulation model developed to model freeways, urban traffic and public transit operations [7]. It performs trajectory-based analysis that utilizes psycho-physical driver behaviour developed by Wiedemann. VISSIM consists of some simulation parameters which have a direct influence on traffic flow, speed and capacity. VISSIM works based on the car-following driver behaviour model developed by Wiedemnann. Some of the parameters have been found to be influential at capacity, while some of them affect the driver's behavior only under car-following situations[7]. The car-following behaviour contains ten different driver-related parameters ranging from CC0 to CC9 with their default values. In different scenarios, researchers examined these parameters' values and tested these parameters' sensitivity for simulated outcomes.

In VISSIM, LC behaviour has default values of specific parameters evaluated for the type of traffic prevailing in Europe and other developed countries[11]. Lane changing models are one in every one of the essential elements of any microscopic traffic simulation. Therefore, developing an additional accurate LC model is a critical requirement for simulating drivers' behaviour. Lane changing manoeuvres was discovered to have a considerable influence on traffic flow characteristics due to the interfering impact they need on surrounding vehicles[12]. The Wiedemann VISSIM model's basic idea assumes that a driver can be in one of four driving modes; Free driving, Approaching, Following and Braking. The driver switches from one mode to another as soon as he/she reaches a certain threshold that can be expressed as a combination of speed difference and distance[13].

The VISSIM model classifies lane changes into two basic categories: free lane change and necessary lane change. In the case of a free lane change, the lane-changing model checks if the available lag time to a collision between the subject vehicle and the following vehicle in the target lane satisfies the "desired safety distance" and "minimum time headway". For a lane change in a queue, the model equally assesses the lead time to collision between the subject and the preceding vehicles in the target lane[14]. These parameters are deemed critical for model LC behavior through VISSIM.

Woody [13] conducted analysis centered on two elements relevant to the calibration and validation process; (a) calibration and validation methods for microsimulation traffic models and (b) adjustment of calibration parameters for freeway models. A sensitivity analysis was conducted on operational calibration parameters in VISSIM, including car following behavior, necessary lane-changing behaviour, and lane changing distances. Results from the sensitivity analysis showed that the headway (CC1), following variation (CC2), and oscillation acceleration (CC7) were the most influential Wiedemann 99 parameters for calibration of mainline freeway segments. In another study, Park [15] developed a procedure for simulation model calibration and validation and demonstrated the proposed method through a case study of coordinated actuated signal system using VISSIM. The study was performed based on a one day data and two measures of performance. After the data analysis, it was discovered that all the multiple runs were not statistically equal to the field distribution. In other words, the simulation output has passed the statistical test at different percentiles for each parameter set.

In a critical review of Liu [16] to determine the sensitive parameters associated with queue length and travel time, VISSIM was employed to examine the sensitive parameters at normal intersections and roundabout. The authors discovered that the average distance between parking, additional parts of the safety distance, multiple parts of the safety distance, the minimum headway are more sensitive parameters on the queue length or travel time at the intersection. Hence, it was suggested that these should be selected as the set of parameters to be corrected at first in the course of calibration. In another previous study. Lownes and Machemehl [17] examined the sensitivity analysis of VISSIM simulation capacity output under various values of driver behaviour parameters. The investigation was undertaken for a simulation developed for the interchange of US-75 and SH-190 north of Dallas, Texas, USA. For each driver behaviour parameter, including look-back distance, simulated capacity variations are noted as the specific parameter modified. Based on the finding, it was recommended that Look-back distance should be carefully calibrated for each particular situation so that the final simulation represents the real-world behaviour of traffic at best in both quantitative and qualitative perspectives.

Most recent, an application of microscopic traffic simulation flow using VISSIM for generating traffic flow data to obtain the essential parameters was reported by Srikanth[5]. The VISSIM model in the study was calibrated and validated based on the field data. Moreover, LC behaviour was analyzed with homogeneous vehicle type traffic on four-lane, six-lane, and eight-lane divided highways segments through VISSIM simulation model. The results revealed that addition of a lane offer more opportunity to the vehicles to change lanes but at the same times causes a reduction in traffic stream speed under non-lane disciple situation, which results in a reduction of per lane capacity [5].

Gomes [18] presented a technique for constructing and calibrating a detailed model of a freeway using VISSIM. The method was applied to a 15-mile stretch of I-210 West in Pasadena, California, USA. The model construction procedure comprised of (a) identification of critical geometric features, (b) collection and processing of traffic data, (c) analysis of the mainline data to identify recurring bottlenecks, (d) VISSIM coding, and (e) calibration based on observations from (c). A qualitative set of goals was established for the calibration process. These were met with relatively few modifications to VISSIM's driver behaviour parameters (CC-parameters).

An extensive review of the existing literature on LC execution, FGD and time headway that were simulated and calibrated in traffic simulation programs showed limited studies regarding drivers' behaviour during LC maneuver through VISSIM. Specifically, there are rare studies relating to Malaysian drivers' behaviour during LC, particularly using an instrumented vehicle, which allowed for the record of real-time relevant traffic events during the data sampling [12]. This paper investigates the FGD on a representative Malaysian expressway and calibrate the data using VISSIM traffic simulation program to explore the subject relating to Malaysian traffic condition. It is expected that this could provide information on Malaysian drivers' behaviour relating to FGD on multilane highways.

Generally, rare studies were conducted on FGD on multilane highways, concerning the estimation of the sensitive parameters using VISSIM simulation approaches as well as its modelling. Few among the related studies [5], [10], [12], [14], [15], [18], [20]worked on the estimation of the calibrating parameters in VISSIM, in which some of them focused on driver's behaviour parameters such as FGD and time headway. As stated earlier, lack of adequate information on FGD regarding Malaysian drivers, particularly relating to their behaviour, necessitated the current investigation.

Berkeley [21] presented a procedure for calibration of default parameters in VISSIM and discovered that driver's behaviour parameters, namely, CC0 (standstill distance) and CC1 (time headway) influences the capacity values. Further, for low-speed conditions, the impact of CC0 on capacity was significant, but as the speed increases, its effect diminishes while the impact of CC1 increases. In similar study, Park [22] described a technique for calibration of default parameters in VISSIM. They established that the safety distance reduction factor defined for the link indicates the aggressiveness of the driver. Sun [23] also performed a parameter calibration of VISSIM using an optimization algorithm approach. The authors developed a new methodology for evaluating the capacity by matching the speed-flow graphs from the field and simulation.

In microscopic traffic simulation, it is usually expected that all calibration parameters within the microscopic simulation model must are identified. Examples of the controllable calibration parameters include; lane change distance, desired speed, and minimum distance headways in the simulation model. Likewise, acceptable ranges for each of the calibration parameters should be well established [18].

Based on the preceding reviewed literature, it is evident that analysis and modelling of LC behavior using traffic simulation approach have been widely used for the past few decades. The application of simulation models is thus essential for solving the complexities associated with modelling macroscopic and microscopic traffic flow behaviours. In relation to traffic simulation application for calibration and validation of traffic parameters, key few researches are discussed as follows. Simulated effects are extremely sensitive to parameters; the literature indicates, namely, CC0 and CC1. As mentioned earlier, the term CC0 is the standstill distance in meters and CC1 is the time headway in seconds between two vehicles at any speed V (m/s). Other parameters CC2 to CC9 have different influence on driver behavior of simulated section[18]. The VISSIM user manual describes CC2 as the actual following distance oscillation process. The parameter CC2 controls the degree to which the driver's distance oscillation is gentle and gradual or sudden and violent. Obviously, parameter values represent a situation between these two extremes, and a small portion of drivers may be defined by these extremes[17]. VISSIM model calculates the safe distance between a following and leading vehicle using Equation 1.

$$\text{Safe distance} = CC0 + CC1 \times \sqrt{V} + CC2 \quad (1)$$

where, CC0, CC1 andCC2 are as defined earlier, V is speed of the leading vehicle (m/s)

Concerning the current study, it is worthy to mention that the investigation in this paper is founded based on the review by Woody [13]. The author focused on three parameters relating to drivers' behaviour; referred to as CC-parameters. These parameters include; CC0 (Standstill Distance), CC1 (Gap Time) and CC2 (Following Distance Oscillation). Hence, the current study also focuses on these three parameters, which examine the Malaysian expressway's condition.

3.0 Methodology

3.1 Field Data Collection

This study's data were collected on a predefined representative segment along Kuala Lumpur – Seremban expressway, Malaysia with approximately 50.2 km length. The section utilized for the study was chosen because it is in good condition with a sizable volume of traffic. It is one of the major links of many parts of Malaysia to the capital city. Equally, the expressway's traffic volume comprises varying vehicles classes with considerable occurrences of lane changing manoeuvres due to the presence of many on- and off-ramps within the selected study section. This could allow for a reasonable sample of lane changing incidences observed on field. The expressway has three lanes in each direction of traffic with an approximate lane width of 3.70 m each with a paved shoulder of 1.50 m. The posted speed along the expressway is mostly 110 km/h. Data were collected during daylight period under varying traffic conditions. for a total of 3 hours on two days typical weekday, the first day data collection between 3:00 to 4:30 PM, the second day data collection duration between 3:30 to 4:30 PM. Data collection was performed in two direction of the chosen section. This study only considers LC incidences involving cars and heavy vehicles as they exhibit varying LC manouevres characteristics. Mainly, heavy vehicles are well known for being associated with low travel speed and usually used the outermost lane; hence, their LC behaviour is certainly different from those of cars. On the other hand, motorcycles' LC manouevres were not considered in this study as different LC manoeuvre's behaviour characterizes them.

The speed of the test and the following vehicles, FGD, traffic volume, and lane changing numbers using an instrument vehicle equipped with VBox. The VBox camera was calibrated for the estimation of the FGD. A regular saloon passenger car instrumented with a Video Velocity Box (VBox) was used for the data collection. A VBox is an on-board real-time data recording system used for observation of traffic events, speed of the instrumented vehicle, acceleration, and lap timing and so forth. It comprises a VBox video recorder, 10Hz GPS data logger, a video camera utilized to record traffic events recording, SD card, and a microphone. In this study, the VBox camera was fixed on the instrumented vehicle, facing backward for recording of LC incidences as the instrumented vehicle traverses the study segment. The VBox system has been established as promising for traffic studies through video recording technique as reported in an earlier study by Woody [13] that utilized it for measuring platooning on two-lane highways. Figure 2 shows a typical view of the segment's part used for the study.

Figure 2. Sample view of traffic on road during field observation

The data was then extracted manually from the video recording played back on the computer screen monitor [7]. The speed of the test vehicle was extracted when the following vehicle started changing the lane. At the same time, the speed of the following vehicle's was measured. Table 1 shows descriptive statistics of the data collected on site.

Table1. Descriptive of data for the two days (volume, LC frequent and speed)

	First day (3:00-4:30 pm)		Second day (3:30-4:30 pm)	
	Test Veh	Following Veh	Test Veh	Following Veh
Volume v/h	--	1225	--	1015
LC Frequency	75		20	
Minimum speed m/s	17.39	18.75	23.96	25.00
Maximum speed m/s	25.56	37.5	28.13	37.5
Average speed m/s	22.41	27.1	26.00	30.3
St.deviation	1.5924	4.3856	1.3625	3.1676

Table 2 illustrates the speed of following vehicle according to vehicle types on two days data collection.

Table 2. Descriptive field speed statistics

	Vehicle Type	Speed Parameters (km/h)				
		Percentage (Numbers)	Maximum	Minimum	Average	Std. Deviation
First day	Car	89.33 (67)	135	67.5	97.24	15.782
	HV	10.66 (8)	135	83.08	100.28	15.758
Second day	Car	90 (18)	135	90	108.84	10.82
	HV	10 (2)	119.7	119.7	119.7	0

For the development of the VISSIM model, the first day data set (75 frequent lane changes) were used for calibration while those of the second day (20 regular lane changing) were utilized to validate FGD in VISSIM.

In the current study, the main variable, the following gap distance (FGD), was estimated as the longitudinal spacing between the rear bumper of the instrumented test vehicle and the front bumper of the immediate vehicle following the former. Table 3 presents the FGD data measured at the field on two days data collection. Figure 3 displays the methodology that followed in this paper.

Table 3. Field measures of FGD (m)

	First day (3:00-4:30 pm)	Second day (3:30-4:30 pm)
Sample size	75	20
Minimum FGD (m)	4.9	4.89
Maximum FDG (m)	50.5	51.21
Average FGD (m)	18.67	18.80
St. Deviation	10.821	12.45

4.0 Calibration Procedures

In the calibration procedures, we first examined the default VISSIM values parameters standstill distances (CC0), minimum headway (CC1) and following distance oscillation (CC2). The selection of these three parameters (CC0, CC1 and CC2) for the homogenous traffic conditions, were conducted based on reviews. The FGD was estimated through VISSIM with default values of all parameters. Table 4 presents the FGD value obtained from the field and FGD value obtained from VISSIM using the default parameters values.

Table 4. Descriptive statistics of FGD (m) obtained from the field and VISSIM

FGD	N	Minimum	Maximum	Mean	Std. Deviation
Field Data	75	4.9	50.5	18.67	10.821
VISSIM Default 10 (runs)	75	8.72	9.02	8.837	0.06

The simulation was conducted for 10 runs and the simulated FGD was obtained. Figure 4 and Figure 5 show the FGD results for observed and simulated data.

As shown in Figure 4 and Figure 5 above, there are huge variations in the FGD for both observed and simulated data. Therefore, the default parameters of the VISSIM model are changed, for further calibration. Table 5 shows the proposed parameters values for each of the selected parameters (Case 1-Case 5).

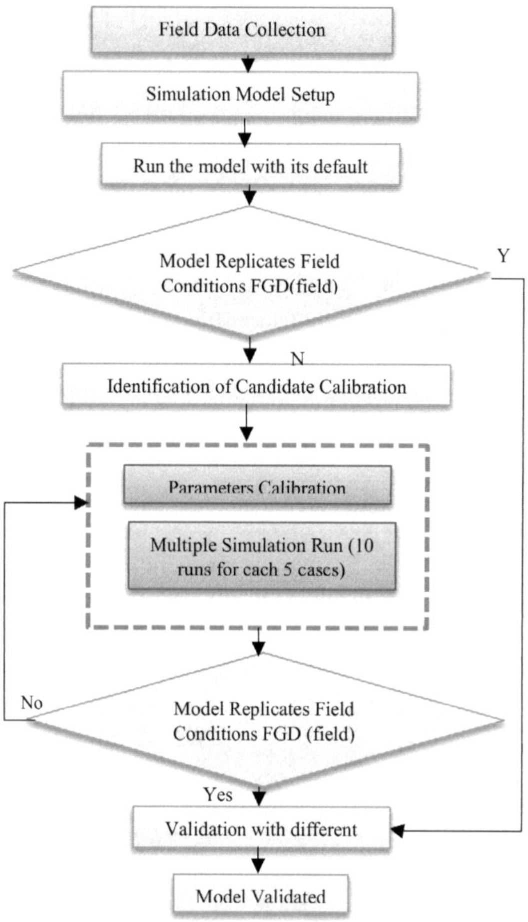

Figure 3. Methodology chart

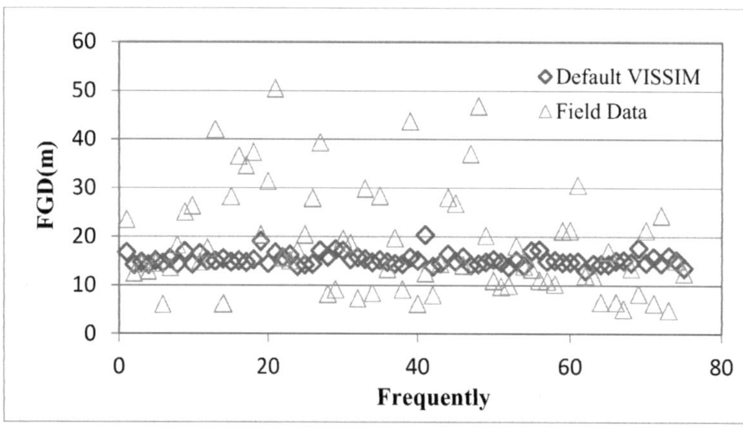

Figure 4. Results for FGD for both field and simulated data

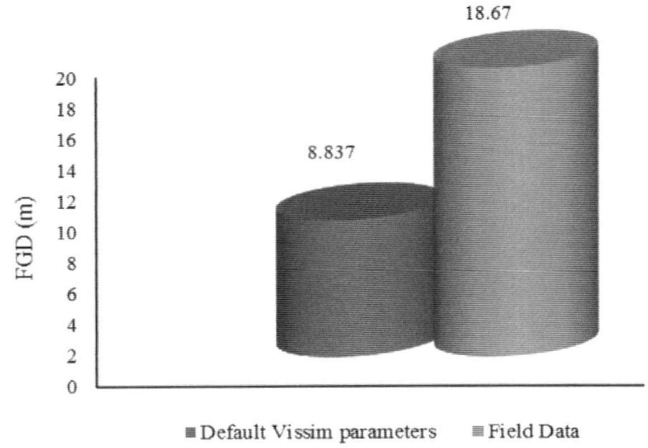

Figure 5. Average FGD for both field and simulated data

The calibration process was ended when a good match between the FGD value obtained between the simulated model and field. In total, five different cases were examined, each case consist of different parameters values as shown in Table 5. Table 6 listed the obtained FGD values from the five cases and the field data. A comparison of the average value of FGD estimated on the field and average value of FGD estimated from each one of the five calibration cases were presented in Figure 6.

Table 5. Candidate Parameter Sets[15]

Parameters	Default	Case 1	Case 2	Case 3	Case 4	Case 5
Standstill Distance CC0 (m)	1.5	1.55	1.60	1.65	1.7	2.50
Minimum Headway CC1(sec)	0.9	0.9	0.9	0.9	0.9	0.9
Following Distance Oscillation CC2(m)	4.0	3.5	4.5	4.55	5.0	8.0

Table 6. Descriptive statistics of FGD(m) of field data and the five cases

FGD	N	Minimum	Maximum	Mean	Std. Deviation
Field Data	75	4.9	50.5	18.67	10.821
Case1	75	13.1	20.24	15.197	1.228
Case2	75	13.43	20.57	15.504	1.298
Case3	75	13.53	20.71	15.614	1.34
Case4	75	13.96	20.85	15.751	1.418
Case5	75	13.23	24.11	17.31	2.28

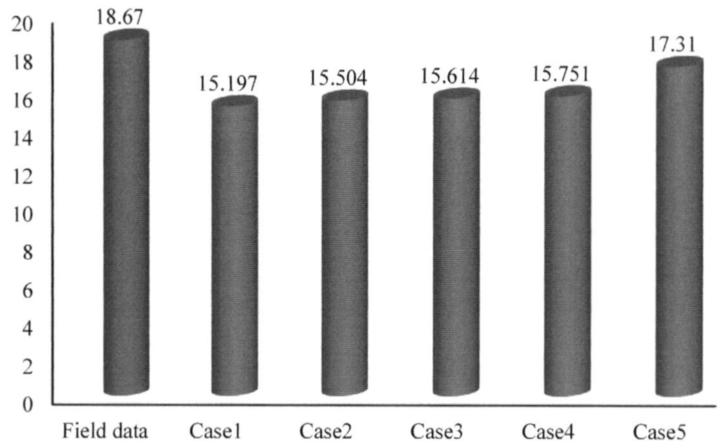

Figure 6. Comparison of FGD from the field and the five cases

4.1 Hypothesis Testing

In order to verify which one from the five cases represent the best calibration result hypothesis tests were conducted to determine if the average FGD obtained from the field is with good agreement with FGD estimated from the model. The test was conducted using 2-tailed t-test at a 5% level of significance ($\alpha = 0.05$) (H_0 null hypothesis: FGD (field data) = FGD (simulation). HA: FGD (field data) $\neq$ FGD (simulation) we reject H_0; null hypothesis if FGD of the simulated case is less than 0.05.

The purpose of the hypothesis testing for multiple scenarios was to determine if any significant difference exists between the means of two groups (a group of field data and cases data). The summary of the main conclusions of the T. test (p-value) process has been presented in Table 7.

Table 7. Significance of default VISSIM values and five cases of comparison with site data

		VISSIM default	Case1	Case2	Case3	Case4	Case5
Field data	p-value	0.0085	5.51E-08	0.0156	0.011	0.0072	0.078
	N	75	75	75	75	75	75

Based on the results listed in Table 7 there is significant difference between the FGD obtained from the field and the simulated FGD estimated from all the cases except case 5 as p-values are lesser than 0.05. Case 5 has p-value greater than 0.05 (0.078 > 0.05), indicating that there is no significant difference between the FGD obtained from field and simulated in case 5. Consequently, the values of CC0, CC1 and CC2 used in case 5 was adopted for this study.

5.0 Model Validation

Using new data set collected at the same segment at different day was used for model validation. Simulation parameters CC0, CC1 and CC2 for homogenous traffic stream were changed to 2.5 m, 0.90 s and 8.0 m, respectively for case 5 (better case) as indicated in Table 5. Table 8 presents the t-test result between the field and validation data. It is clear that from the t-test result that the model was successfully calibrated and validated.

Table 8. Significance of field data for 2nd day and case5 parameters

		Case5
Field data(2nd day data)	T. test value	0.198
	N	20

6.0 Conclusions

This paper has outlined a methodology for the calibration and validation of simulation model for an expressway. The procedure included the generation and processing of field data from a representative segment on the expressway between Kuala Lumpur and Seremban, Malaysia. The study utilized a two-day data on following gap distance (FGD) as measure of performance. Model parameters such as Standstill Distance (CC0), Gap Time (CC1) and Following Distance Oscillation (CC2) were used as calibration parameters. As established earlier five different cases were examined each case consists of different values of CC0, CC1, and CC2. It is found that the combination of the parameter values in case 5 produced the closest FGD of the field. This result was verified by using other data set collected on a different day. This finding is well consistent with what was established in the existing literature as reported in[13]. Generally, findings deduced from this study could help to identify and quantify the influence of significant traffic parameters on lane changing for safety gap. Further, the research can be performed in VISSIM model to explore whether changing the significant traffic parameters with diverse distributions for a higher number of lanes has a significant positive effect on lane changing as equally suggested by [8]. This study showed that for such expressway that involves complex interactions between vehicles, the VISSIM simulation environment is reasonably suited. It is essential to state that this analysis's results were drawn based on a preliminary study, so it is recommended that more considerable sample data be collected over several days, particularly field data variability. It will also be useful to decide whether the methodology applies to an entire network or just a particular one.

References

[1] G.-L. Chang and Y.-M. Kao, 1991"An empirical investigation of macroscopic lane-changing characteristics on uncongested multilane freeways," *Transp. Res. Part A Gen.*, vol. **25**, no. 6, pp. 375–389.

[2] M. E. Sanik, Prasetijo, F. Mustakim, N. B. Hamid, M. Nor, A. Hakimi, M. Safberi., 2016 "Drivers Lane Changing Behavior at Urban Intersection by Using Gap Acceptance Approach," *ARPN Journal of Engineering and Applied Sciences, vol.* **11**, no. 24, pp. 14070–14074.

[3] H. Wei, E. Meyer, J. Lee, and C. Feng,2000"Characterizing and Modeling Observed Lane-Changing Behavior: Lane-Vehicle-Based Microscopic Simulation on Urban Street Network," *Transp. Res. Rec. J. Transp. Res. Board*, vol. **1710**, no. 0, pp. 104–113.

[4] X. Cao, W. Young, and M. Sarvi, 2015 "A study of traffic conflict with surrounding traffic during mandatory lane-changing execution," in *Australasian Transport Research Forum* .

[5] S. Srikanth, A. Mehar, and K. G. N. V. Praveen, 2020"Simulation of traffic flow to analyze lane changes on multi-lane highways under non-lane discipline," *Period. Polytech. Transp. Eng.*, vol. **48**, no. 2, pp. 109–116,

[6] M. Shoraka, 2016"Development of Simulation Model for Assessing the Performance of Weaving Sections on Interchanges," *PhD thesis*, Universiti Teknologi Malaysia.pp. 1-301,

[7] S. Srikanth, A. Mehar, and A. Parihar, 2017 "Calibration of vissim model for multilane highways using speed flow curves," *Civ. Eng. J.*, no. 3.
[8] D. Farooq and J. Juhasz, 2019 "Simulation-based analysis of the effect of significant traffic parameters on lane changing for driving logic 'cautious' on a freeway," *Sustainability*, vol. **11**, no. 21, p. 5976.
[9] L. Weber, 2017 "Driver Modeling and Simulation of Lane Change Situations: Influence of Different Rear View Mirror Types on Gap Acceptance Behavior," *PhD thesis*, Universität Oldenburg.
[10] W. Xiaorui and Y. Hongxu, 2013 "A Lane Change Model with the Consideration of Car Following Behavior," *Procedia - Soc. Behav. Sci.*, vol. **96**, no. Cictp, pp. 2354–2361.
[11] A. Mehar, S. Chandra, and S. Velmurugan, 2014 "Highway capacity through VISSIM calibrated for mixed traffic conditions," *KSCE J. Civ. Eng.*, vol. **18**, no. 2, pp. 639–645.
[12] S. Moridpour and G. Rose, 2010 "Modeling the Lane Changing Execution of Multi Class Vehicles under Heavy Traffic Conditions," *Transp. Res. Rec. J. Transp. Res. Board*, pp.2-26
[13] T. Woody, 2006 "Calibrating freeway simulation models in VISSIM," *Master thesis Univ. Washingt.*.
[14] E. Balal, R. L. Cheu, and T. Gyan-sarkodie,2014 "Analysis of Discretionary Lane Changing Parameters on Freeways," *Int. J. Transp. Sci. Technol.*, vol. **3**, no. 3, pp. 277–296.
[15] B. Park, J. Won, and I. Yun,2006 "Application of microscopic simulation model calibration and validation procedure: Case study of coordinated actuated signal system," *Transp. Res. Rec.*, vol. **1978**, no. 1, pp. 113–122.
[16] W. W. Liu, Y. Qin, H. H. Dong, and Y. F. Yang, 2014 "Driving behavior parameter sensitivity analysis based on VISSIM," in *Applied Mechanics and Materials*, vol. **668**, pp. 1453–1457.
[17] N. E. Lownes and R. B. Machemehl, 2006 "Sensitivity of simulated capacity to modification of VISSIM driver behavior parameters," *Transp. Res. Rec.*, vol. **1988**, no. 1, pp. 102–110.
[18] G. Gomes, A. May, and R. Horowitz,2004 "Congested freeway microsimulation model using VISSIM," *Transp. Res. Rec.*, vol. **1876**, no. 1, pp. 71–81.
[19] S. Srikanth, A. Mehar, K. Guru, and N. Venkata,2017 "Simulation of Traffic Flow to Analyze Lane Changes on Multi-lane Highways under Non-lane Discipline," *Period. Polytech. Transp. Eng.* pp. 1–7.
[20] S. Datta, 2018 "Assessment of uncontrolled intersections through calibration of VISSIM for Indian Traffic Conditions," in *Industry Interactive Innovations in Science, Engineering and Technology*, Springer, pp. 323–338.
[21] G. Gomes and R. Horowitz, 2004 "Calibration of VISSIM for a Congested Freeway," *Res. Reports, Calif. Partners Adv. Transit Highw. (PATH), Inst. Transp. Stud. UC Berkeley*.pp.2-48,
[22] B. B. Park, D. Ph, and J. Won, 2006 "Simulation Model Calibration and Validation : Phase II : Development of Implementation Handbook and Short Courses," *Virginia Transp. Res. Counc.*pp.5-28,
[23] S. Menneni, C. Sun, and P. Vortisch, 2008 "Microsimulation calibration using speed-flow relationships," *Transp. Res. Rec.*, vol. **2088**, no. 1, pp. 1–9.

IOP Conf. Series: Materials Science and Engineering 1153 (2021) 012017 doi:10.1088/1757-899X/1153/1/012017

Investigation of Pile Behaviour using Rotary-Jacking Method

N S Hassan[1] and A Ibrahim[2]
[1]Centre for Graduate Studies, Universiti Pertahanan Nasional Malaysia, Kem Sungai Besi, 57000 Kuala Lumpur.
[2]Civil Engineering Department, Faculty of Engineering, Universiti Pertahanan Nasional Malaysia, Kem Sungai Besi, 57000 Kuala Lumpur.
Email: aniza@upnm.edu.my

Abstract. This research investigates the installation of a new pile using the rotary and jacking method. The use of hydraulic ram to apply static jacking force does not produce the noise and ground vibration produced by the traditional percussion and vibro-hammer methods. The pile jacking process can increase the bearing capacity and stiff base response but may occasionally cause excessive installation resistance. An innovative rotary jacking method can overcome these problems and enhance pile performance. The research seeks to determine the effect of pile loading and the pile response under axial loading. The laboratory experiment uses a 25-mm circular pile and a screw pile together with a large container filled with silica sand attached to a rotary jacking machine. The installed piles are subjected to the installation and static load tests. The results show that the rotary jacking method successfully installed the pile with less resistance between the sand and pile surface, which allows the piles to penetrate the required depth with less loading. The experimental results proved that the rotary jacking method reduced soil settlement and the time taken to drive the pile into the soil. This new innovative method will enhance pile installation in the construction industry.

1. Introduction

Pile driving is an essential part of building constructions. Piles transfer the load from structures to the ground and strengthen the structure foundation. Pile foundation is defined as installation of pile to transfer the load to the ground when upper structure is not strong enough due to soil weakness [1]. Two types of foundation are classified which are deep and shallow foundation. Deep foundation is widely used because its large bearing capacity, low distortion, and high reliability [2]. The cost of building a foundation on loose soil is very high, particularly in urban areas. Piles provide support to the retaining wall, bridge piers, and machinery foundations which carry the combined vertical and horizontal loads [3]. Pile driving into the ground of a construction site generates vibration as the pile penetrates the soil. Even though the natural vibration of the earth and human activities generate ground vibration [4], the vibration generated during pile driving can cause discomfort to people and damage the nearby structures [5]. Some pile installations, such as hammering installation, have an adverse impact on the environment. The overload force exerted during pile installation may cause underground cracking and damage the water and sewer pipes, power lines, and communications cables [6]. The current pile installation methods take longer to drive the pile into the ground. Longer construction time means higher costs. Malaysia has been using a jack-in piling with a working load of up to 3000 kN since the late 1990s [7]. Jacking method is the most familiar method used at construction site to install the pile into the ground.

It is essential to use a suitable pile for a structure to ensure structural stiffness. The type of pile used is dependent on several factors, including soil type, corrosion, local availability and cost, contractor preference, and the load-bearing requirement of the foundation [8]. The friction between the soil and pile surface during pile installation can produce resistance. Pile resistance is one of the factors that affecting the performance of pile behaviour. Previous study states that, the ground resistance of single pile is determined by stress distribution under pile base and skin surface [9]. Stainless steel piles are used in construction work. The friction generated between the steel and soil may exceed 35-50% of the total draft resistance [10].

The rotary jacking method is an innovative method for installing pile in the ground developed from the traditional axial jacking method. It uses rotation movement to drive the pile to a particular depth. During the installation, the pile is simultaneously rotated and jacked into the ground. Pile installation by rotation movement reduces the required axial loading [11]. However, the pile response under subsequent loading of the jacking and rotation method is not yet understood.

This paper seeks to determine the impact of using different pile installation methods, the resistance of different pile shapes and sizes, and the subsequent response under axial working loading by rotary jacking method.

2. Materials and Method

2.1. Materials

This experiment used a 25-mm diameter and 600-mm long stainless-steel pile and screw pile, as shown in figure 1 and figure 2. The silica sand passed a 600-µm sieve but retained in a 300-µm sieve. The sand was compacted for ten minutes to reduce the pore volume containing water or air. The piles were driven into the silica sand. The experiment followed the BS 1377: PART 2:1990: Clause 9.2 [12].

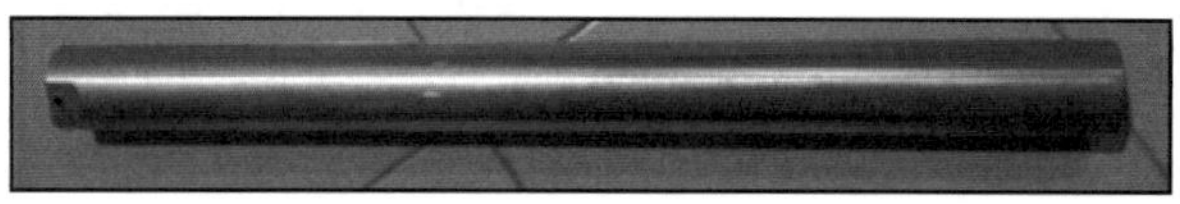

Figure 1. 25 mm circular pile

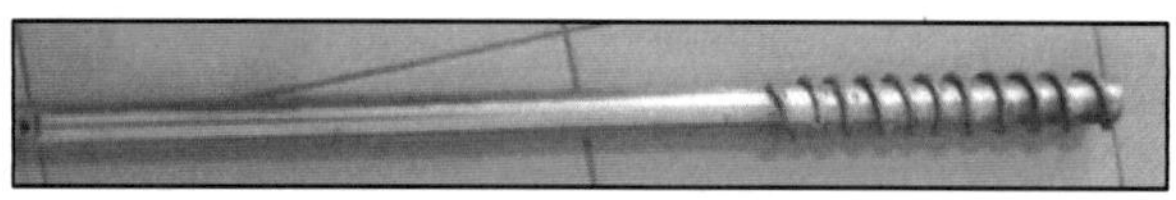

Figure 2. screw pile

2.2. Apparatus Set-up

2.2.1 Custom Rotary Jacking Machine. The rotary jacking machine in this experiment comprises several components, as shown in figure 3. The custom machine is 0.8 meters high and 0.8 metre in diameter and has a large, circular container to hold the silica sand. The rotary jacking machine has a control panel with three operation switches for penetration, rotation, and vibration. The penetration and rotation switch controls two pile movements, forward and reverse. The machine also connects directly to the data logger that collects the data.

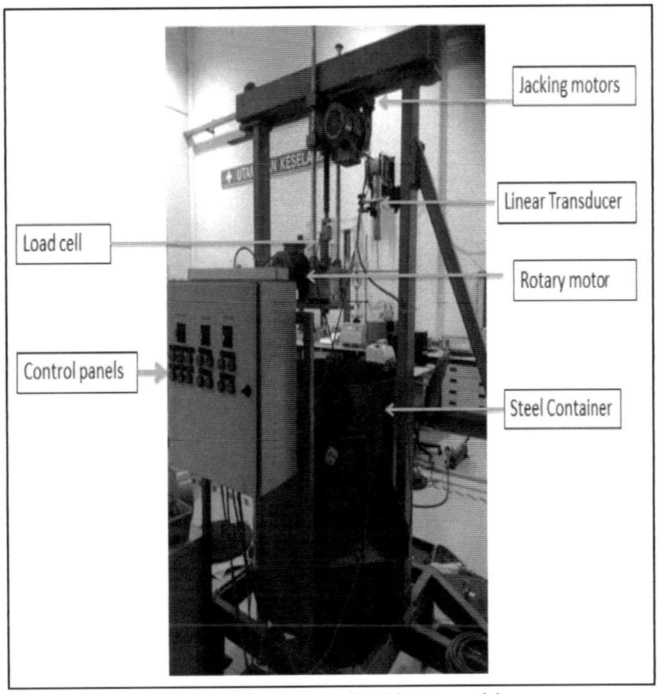

Figure 3. Rotary-jacking machine

2.2.2 Data logger. This research used the Campbell Scientific Datalogger (CR800) to measure the electrical signals and convert the measurements into engineering units. The data logger is connected to the Loggernet software used to analyse the collected data.

2.2.3 Motors. Motors. The mechanism system comprises two (2) motors, the jacking motor and the rotating motor. The maximum speed of the rotating motor is 1 min/rev and 0.0043mm/s for the jacking motor.

2.2.4 Load Cell. The load cell is a transducer for producing an electrical signal that measures the magnitude directly proportional to the force. The maximum capacity of the load cell is 2000 pounds, which is equivalent to 8.90 kN. The vertical pile displacement was measured using the linear transducer, and the data was displayed on the computer.

2.3. Experiment

The experiment consists of pile installation and static loading testing. The jacking method and the jacking and rotary methods were used to install the pile. The first step is vibrating the silica sand in the container for five minutes to compact the sand. The first pile installation used the jacking method and took about 2.5 hours to reach the required length. Because the machine cannot be driven to 600-mm pile length, the maximum length of pile installation is 300mm. The static load test used a 0.08 kN weighing block. Because the machine was custom-made, the highest loading is ten blocks weighing 0.78 kN. The data was recorded immediately after the completion of the pile installation. The pile was removed after the static load test, and the second pile was installed using the same process. Table 1 lists the experiments.

Table 1. Experimental schedule

Shape of Pile	Method of Installation		Static Load Test	
	Jacking	Jacking and Rotary	Jacking	Jacking and Rotary
25 mm Circular	√	√	√	√
Screw Pile	√	√	√	√

3. Result and Discussion.

3.1. Installation Method

The pile installation used the jacking method and the jacking and rotary method. Figure 4 shows the graph for installing the 25-mm circular pile. The jacking method required 0.60 kN loading to drive the pile to a depth of 150 mm, and the jacking and rotary method required 0.30 kN loading. Figure 5 shows the graph for the installation of the screw pile using both installation methods. It shows that the jacking method required 1.03 kN loading, and the jacking and rotary method required 0.66 kN loading to drive the pile to a depth of 150 mm. The loading required for the jacking method is twice that for the jacking and rotary methods, which proves that the jacking and rotary method required less loading to drive the pile to a particular depth. The lower loading means less resistance between the pile surface and sand, which reduced the time for installing the pile in the sand.

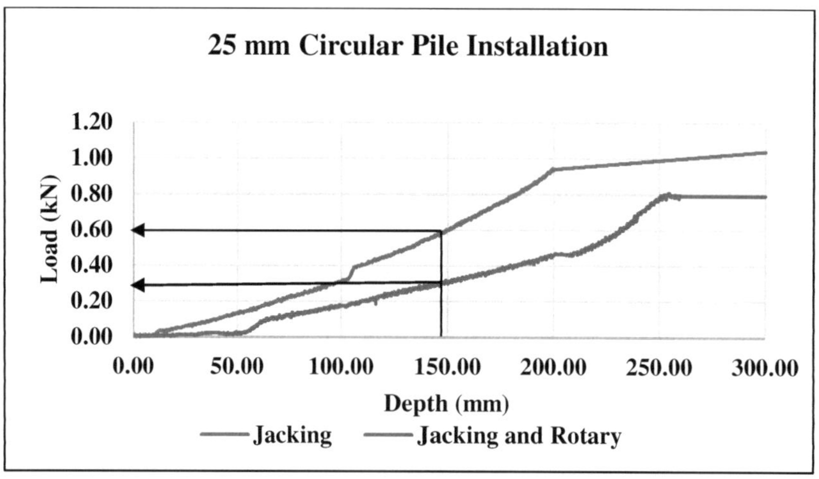

Figure 4. Result of 25 mm circular pile installation.

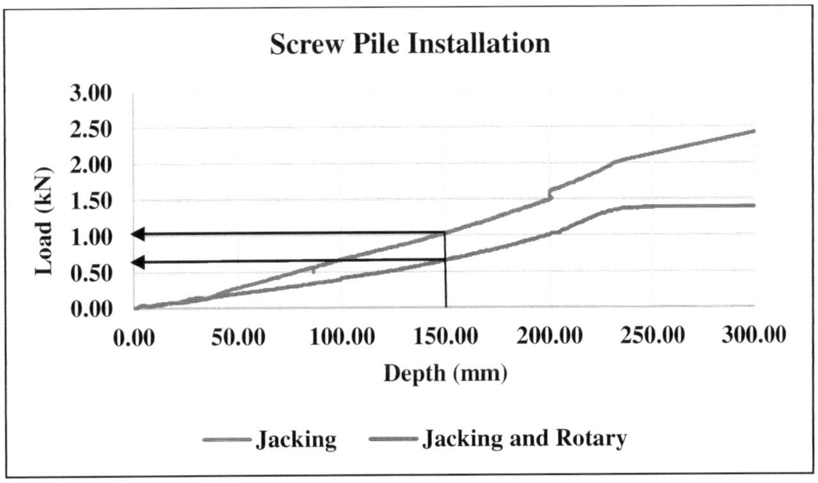

Figure 5. Result of screw pile installation.

3.2. Static Load Test

The static load test determines the relationship between the load and the pile head displacement [13]. This test was carried out to determine the subsequent response that influences the pile settlement. The linear transducer collects the data under working axial loading. The maximum loading in this experiment is 0.785 kN, and the test is carried out immediately after the completion of the installation. Each test takes about one hour. Figure 6 and figure 7 show the static load test results for the 25-mm circular and screw pile for both installation methods. The 25-mm circular pile shows a 1.4 mm settlement at a point load of 0.6 kN when using the jacking method and a 0.4 mm settlement when using the jacking and rotary method. The screw pile showed 1.02 mm and 0.75 mm settlement for the jacking and jacking rotary method at the same point of load. The graphs show less pile settlement for the jacking and rotary method relative to the jacking method. Therefore, the jacking and rotary method resulted in less structural failure and is safer for the surrounding environment.

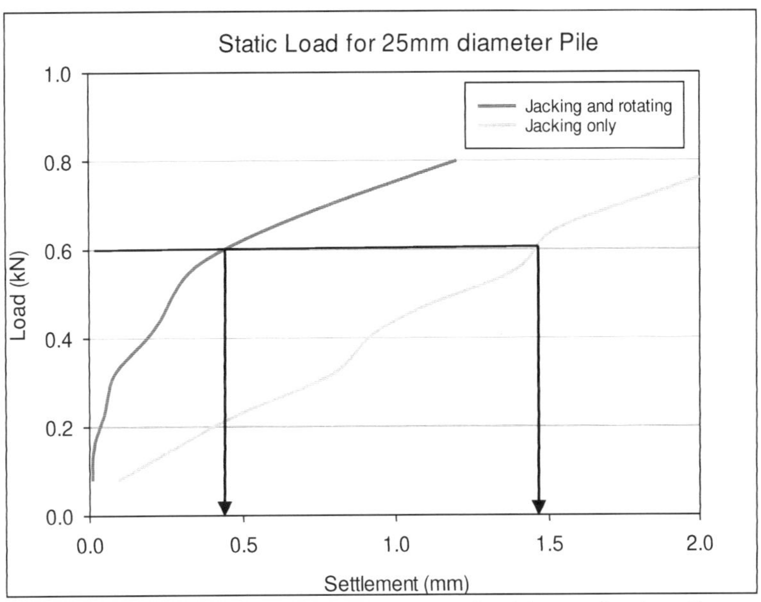

Figure 6. Result of static load test for 25 mm circular pile.

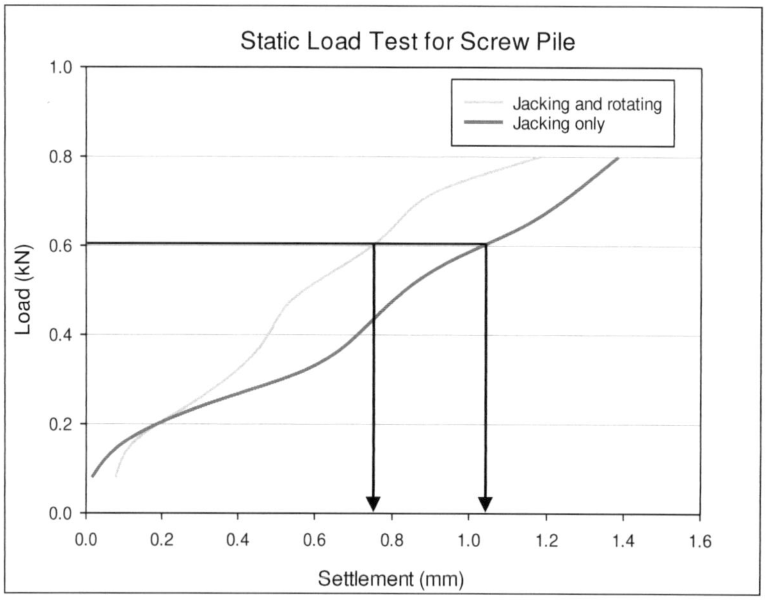

Figure 7. Result of static load test for screw pile.

4. Conclusion

The experimental results showed that the jack and rotary pile installation reduced the friction between the pile and sand, which reduced the resistance. In terms of resistance, bearing capacity, and load carried by the pile (load test), the jacking and rotary method is superior to the jacking method. The time taken for pile installation using the jacking and rotary method is less than for the jacking method. Therefore, the jacking and rotary method is an efficient method for pile installation because it is more eco-friendly and less costly. Future research should use different types of sand to determine the effectiveness of the jack and rotary method on other soil parameters. Also, further research should seek to improve the results of the installation test.

References

[1] Kim S, Whang, S W and Kim S 2017 Pile Foundations Design Through the Increased Bearing Capacity of Extended End Pile *Journal of Asian Architecture and Building Engineering*, **16** pp 395-402

[2] Zhanfang H, Xiaohong B, Chao Y, and Yanping W 2020 Vertical Bearing Capacity of a Pile-Liquefiable Sandy Soil Foundation Under Horizontal Seismic Force *Plos One*, **15**

[3] Letsios C, Lagaros N D and Papadrakakis M 2014 Optimum design methodology for pile foundation in london *Case Study in Structural Engineering.* **2** pp 24-32

[4] Athanasopoulos G A and Pelekis P C 2000 Ground vibrations from sheet pile driving in urban environment: measurements, analysis and effects on buildings and occupants *Soil Dynamic and Earthquake Engineering*. **19** pp 371-387

[5] White D, Finlay T, Bolton M and Bearss G 2002 Press-in piling: Ground vibration and noise during installation *Cambridge University Department.* **116** pp 363-371

[6] Endres D 2013 Effect of pile driving induced vibrations on nearby structures *Geotechnical Services Section Michigan Department of Transportation*. RC-1600

[7] Tan. Y C, Chow C M and Gue S S 2011 Jack-in pile in Malaysia: A Malaysia consultant respective *Proceedings of the VLA Seminar.*

[8] Adejumo T E and Boiko I L 2012 Pile Foundation and Installation Methods in Nigeria *Faculty*

IOP Conf. Series: Materials Science and Engineering 1153 (2021) 012017 doi:10.1088/1757-899X/1153/1/012017

of Civil Engineering. **17** pp 2613-2621
[9] Norkus A and Martinkus V 2019 Experimental Study on Bearing Capacity Resistance of Short Displacement Pile Groups in Dense Sands. *Journals of Civil Engineering and Management*, **25** pp 551-558
[10] Vilde A, Rucins A and Sevostjanovs G 2007 Impact of speed on the soil sliding resistance. *Jelgava* **5** pp 24-25
[11] Hazla E 2013 Rotary press-in piling hard ground. *Fourth-Year Undergraduate Project in Group D.*
[12] British Standard (BS) 1377: PART 2:1990: Clause 9.2 Methods of test for soils for civil engineering purposes
[13] Krasinski A and Wisaniew M 2017 Static load test on instrumented pile-field data and numerical simulations. *Studia Geotechnica et Mechanica,* **39**

IOP Conf. Series: Materials Science and Engineering 1153 (2021) 012018 doi:10.1088/1757-899X/1153/1/012018

Critical micelle concentration, interfacial tension and wettability alteration study on the surface of paraffin oil-wet sandstone using saponin

A H Alpandi[1,2*], F A Inasyah[2], A Sidek[2], H Husin[1], R Junin[2] and M Z Jaafar[2]

[1]Department of Petroleum Engineering, Universiti Teknologi Petronas, 32610 Seri Iskandar, Perak, Malaysia.
[2]School of Chemical & Energy Engineering, Faculty of Engineering, Universiti Teknologi Malaysia, 81310 Skudai, Johor Bahru, Malaysia.

*Corresponding author: amni_17009776@utp.edu.my

Abstract. Nowadays, the implementation of chemical enhanced oil recovery (CEOR) techniques improves the oil recovery by injecting synthetic surfactants in reservoir. However, the recent use of this synthetic surfactant is quite expensive and possesses toxicity problems when exposed to the environment. This issue has forced the search for alternative, cheaper and natural synthetic surfactant to enhance oil recovery. Hence, this study aims to provide some insights into the effect of saponin as natural plant-based non-ionic surfactant on the surface of paraffin oil-wet sandstone. The effectiveness of saponin depends on the critical micelle concentration (CMC), interfacial tension (IFT) and wettability alteration. The surfactant concentration containing saponin was varying from 0.005wt% to 0.07wt% to determine CMC value using surface tension measurement, meanwhile for IFT measurement was conducted at 0.5wt% to 8wt% concentration using Krus Tensiometer K6. The experiments were conducted at standard condition with 25°C and 14.7psia. The findings showed that surface tension for CMC value of saponin is 0.05wt%. Approximately 27% reduction from initial IFT was achieved with 8wt% of saponin concentration. The saponin successfully alter the wetting state of oil-wet sandstone to intermediate-wet through quantitatively measurement of contact angle. Saponin with low CMC value is favourable to become a reference concentration as natural surfactant, effective to alter of reservoir rock properties, facilitate oil mobilization, increase oil recovery and suggested as an additive for mitigating wax deposition.

1. Introduction

Enhanced oil recovery (EOR) is one of the recent techniques used to satisfy the world's energy needs. However, the main obstacles faced by industry are environmental and economic issues caused by the uses of chemicals in the application of EOR [1]. Hence, the application of chemical EOR which widely used in the most of country especially in US began to decrease significantly starting from 1988. The increasing price of surfactants cause the declination use of this type of chemical. The long-term uses of surfactant also contribute to the environmental pollution. Large volume of synthetic surfactants needed during the application of EOR which affect the increment of raw material. Thus, this condition makes the surfactants no longer economical. Furthermore, synthetic surfactant has high adsorption rate which will render the flooding process inefficient [2].

In oil and gas industry, chemical enhanced oil recovery (CEOR) technique is usually applied to brownfield reservoir. The main purpose of CEOR is to increase the oil mobility. Surfactant is a main material to change the physicochemical properties in aqueous injected fluid because of its ability to reduce interfacial tension (IFT) between water-oil phase. In addition, the surfactants used in the low concentrations may change the wettability of reservoir rock to be intermediate wet or more water wet [3]. Baturaja formation in the May fields located in Indonesia, Mauddud fields in Bahrain, Cottonwood Creek fields in Wyoming, Yates fields and the Cretaceous Upper Edwards reservoir in Texas are oil drilling fields that have been achieved successful results from the application of surfactants.

There are recent researchers conducted research on natural surfactants application as an alternative to synthetic surfactants [4,5,6,7,8,9]. Natural surfactants are more environmental-friendly because of their relatively low toxicity and economically cheaper than synthetic surfactants. The utility of natural surfactants provides various benefits such as simple chemical structures which become biodegradability, environmental compatibility, composed of specific functional groups thereby increasing selectivity, detoxification of specific pollutants which allow for specificity, activity in extreme temperatures, certain pH and salinity. In general, the role of natural surfactants is to increase the spread of contaminants in the aqueous phase and increase the bioavailability of hydrophobic substrates for microorganisms, in order to eliminate pollutants through biodegradation [10].

The main source of plant-based surfactants that have been widely used in various industries is saponin [11,12,13,14]. The saponin is a non-ionic natural surfactant which consist of a non-volatile glycoside compound with a high molecular weight. It is composed of sugars and aglycones which relate to hydrocarbon groups. In oil and gas industry, saponins are used as an agent in CEOR with the same function as the synthetic surfactants application process. Figure 1 illustrated the general molecular structure of saponin.

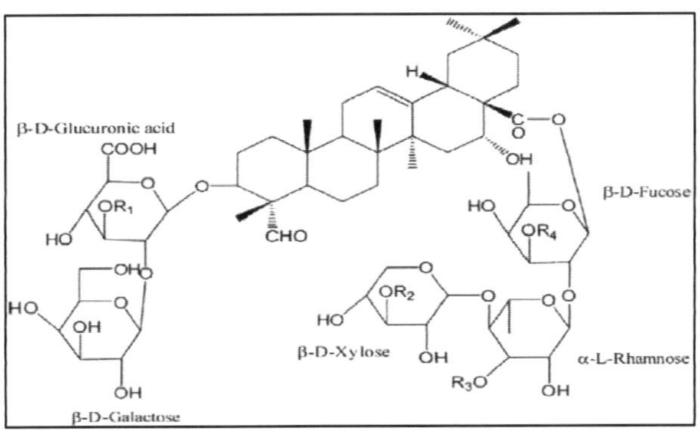

Figure 1. General molecular structure of saponin [12].

The aim of this study is to determine the critical micelle concentration (CMC) of saponin based on surface tension measurement from Du Noüy ring method. Besides that, this study was conducted to identify the effect of saponin on reservoir rock and fluid properties which include the IFT measurement from the same method as conducted in surface tension measurement. The wettability alteration from quantitatively evaluation in contact angle measurement by implementation of sessile drop method also has been discovered. In addition, oil-wet sandstone was used as porous media in this study as about 61% of giant fields in the world is sandstone reservoir.

Many researchers conducted a CMC measurement of saponin extracted from various type of plant by using surface tension as a parameter. According to Chen et al. [13], the additional of extracted saponins from *Quillaja saponaria* with 13.9% purity gives the CMC value of saponin which ranging from 0.01-0.02 wt% at neutral pH of 6.5 s.u. Meanwhile, Stanimirova et al. [12] conducted the same experiment from the same source of saponin, but the CMC value obtained was 0.025 wt%.

As the natural surfactant, saponin could reduce IFT of water-oil phase which considered as a factor that supports the use of saponins in the CEOR technique. Many researchers have successfully implemented the application of saponin in CEOR method throughout their studies. Shahri et al. [14] conducted a study in IFT reduction and wettability alteration to ensure the effects of saponin addition in distilled water as the surfactant solution. The results shows that saponin able to reduce IFT from 48 to 9 dynes/cm which makes saponin has similar ability as conventional surfactants that commonly used in chemical flooding process. The average contact angle of 50.24° was obtained which is in a relatively hydrophilic range. Since IFT is affected by capillary pressure, this pressure governs the wetting changes due to its dependency on contact angle. In addition, IFT reduction between water-oil assist to mobilize oil which can reduce the tendency of oil from stick on the rock grains. Moradi et al. [15] stated that wettability alteration on oil-wet quartz identified by using saponins extracted from *Tribulus terrestris* which indicated by the reduction of contact angle from 160.3° to 80.6° with 0.9 wt% saponin concentration.

2. Materials and Methods

2.1. Materials
The natural nonionic surfactant used from saponin extracted from bitter leaf was supplied by Sigma Aldrich located at Selangor, Malaysia. This surfactant was in powder form with the total amount of 11 grams used to prepare the surfactant solution. In this study, distilled water was used as the solvent in solution preparation. For the evaluation of wettability alteration, sandstone plate from Mount Santubong, Sarawak was used as a porous media with diameter of 50 mm and thickness of 9 mm. In addition, paraffin oil was used as an oil-phase. After aging process, the sandstone was cleaned out with toluene before conducting wettability alteration mechanism and n-heptane was used for washing the sandstone plate.

2.2. Preparation of surfactant solution
Surfactant solution is a mixture of distilled water and saponin powder at different concentrations. This process started by preparing the stock solutions due to minimize the utility of saponin powder. 8 wt% is the concentration for stock solution by mixing 8 grams of saponin powder into 100 ml distilled water. Furthermore, lower concentrations prepared through the dilution process which is carried out by adding distilled water according to the desired concentration based on the equivalent of the concentration and volume of stock solution to the additional volume of concentration. Surfactant solutions with concentrations of 6 wt%, 3 wt%, 1 wt%, 0.1 wt%, 0.07 wt%, 0.06 wt%, 0.05 wt%, 0.03 wt%, 0.01 wt%, and 0.005 wt% were prepared in this study.

2.3. CMC measurement
Du Noüy ring and Krüss Tensiometer K6 used to measure the surface tension at ambient condition with temperature of 25°C and pressure at 14.7 psi [16]. The effect of surfactant solution towards the surface tension was measured by using tensiometer. Afterwards, the result of surface tension at different concentrations were plotted to determine the CMC value. The concentration of surfactant solutions by various amount of saponin that has been used are 0.005 wt%, 0.01 wt%, 0.03 wt%, 0.05 wt%, 0.06 wt%, and 0.07 wt% which is carried out alternately from low to higher concentrations.

2.4. IFT measurement
The experiment was conducted at ambient condition with temperature of 25°C and pressure at 14.7 psi. The concentration of surfactant solution by various amount of saponin was used with 0.1 wt%, 0.5 wt%, 1 wt%, 3 wt%, 5 wt%, 6 wt%, and 8 wt% which is carried out alternately from low to higher concentrations. Identical apparatus was used as for surface tension measurement with all similar

instrument preparation and procedure. The difference in this experiment is only in the use of paraffin oil and surfactant solution. The solution surfactant was poured into a sample container and followed by pouring paraffin oil. The hooked platinum ring was later submerged inside surfactant solution for about 5 mm deep by moving the movable platform upward. As the platinum ring detached from the liquid interface, the IFT between surfactant solution and paraffin oil was analysed which can be directly recorded [1].

2.5. Contact angle measurement

Sessile drop method was implemented in this contact angle measurement which must be initially prepare the oil-wet sandstone through aging process. Initially, the sandstone core was cleaned with toluene to remove fatty acid in sandstone plate which heated at 60°C for 24 hours and the core must be dried at 40°C for 24 hours. The aging process of sandstone in paraffin oil was performed by immersing the plate in paraffin oil for 2 weeks at 60°C and continued for 30 days at ambient temperature which is 25°C in order to induce the adsorption of carboxylic component on the rock surface [15]. The sandstone plate after aging process must be washed with n-heptane and distilled water, then heated in the oven at 60°C for at least 3 hours to be ready for wettability assessment and used in evaluation of wettability alteration as an oil-wet rock.

After the preparation of oil-wet sandstone, this sandstone plate was immersed in 0.05 wt% (according to its CMC value) of surfactant solution for 10 days at 25°C and 14.7 psia, then the measurement was taken for every day by sessile drop method. The paraffin oil was injected to form a droplet on the sandstone plate immersed in surfactant solution and side image of oil drops were captured using digital camera Nikon D90 and Nikon 60 mm micro lens which the angle was analysed from Image software [17,18,19].

3. Results and Discussion

3.1. Critical micelle concentration (CMC)

In this study, surface tension measurement was conducted in the air-surfactant solution at different concentrations with ambient conditions of 25°C and 14.7 psia. The addition of concentrations in saponins succeeded in reducing the value of the initial surface tension from the air-water. The surface tension reduction occurs during the reduction of contraction force. The molecules in the water are replaced by the saponin molecule when adsorption occurs by the saponin molecule on the water surface. This condition causing the attraction forces between water and saponin molecules lower than two water molecules represented by the hydrophobic tail of the skewed skew into the air and the hydrophilic head of saponin leads to water as a surface anchor [20]. This phenomenon will further drop until the minimum surface tension by increasing of saponin concentration. Figure 2 shows surface tension against surfactant concentration with the addition of saponin.

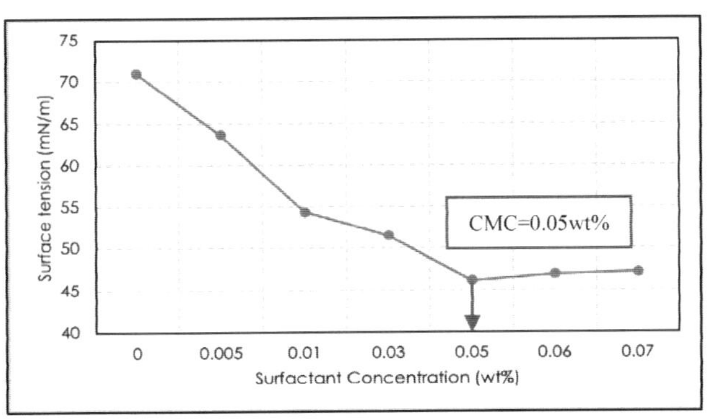

Figure 2. Surface tension against surfactant concentration with the addition of saponin.

Based on figure 2, initial surface tension of surfactant flooding starts from 71 mN/m which significantly decrease to 54.4 mN/m at saponin's concentration of 0.01 wt% and decreases continuously until it reaches the minimum surface tension of 46.13 mN/m at 0.05 wt% of saponin's concentration. The addition of saponins at the minimum surface tension consider the CMC of saponin. Addition of saponin concentration after CMC value shows an increase with an insignificant value on surface tension changes, therefore CMC of saponin is recommended as the most efficient concentration of surfactant in CEOR. Thus, the CMC of saponin is 0.05 wt% which indicated as a minimum surface tension and concentration on the break points from slope of -268.5 (before CMC) which turns into the slope of 52 (after exceeds CMC) as the indicator to be almost constant value of surface tension. This trend changes likely to occur because some of the adsorbed saponin molecules are distorted from the interface to the bulk phase.

The results of this study are insignificantly difference with some previous studies using surface tension as a parameter to determine the CMC of saponins. Chen et al. [13] shows13.9% purity obtained the CMC value of 0.01-0.02 wt% at near neutral pH of 6.5 s.u. Meanwhile, Stanimirova et al. [12] use the same source of saponin observed at 25°C acquired the CMC value of 0.025 wt%. Moreover, Zhou et al. [11] performed an experimental study using conventional saponins from manufacture which identified 0.003-0.015 wt% of CMC value. Another CMC value of saponins from three different manufacturers (Sigma Chemical Co. in St. Louis, MO; Acros Organics in Fair Lawn, NJ; and Penco of Lyndhurst Inc. in Lyndhurst, NJ) is ranged from 0.051-0.072 wt% at 25°C studied by Mitra and Dungan [21].

3.2. Interfacial tension (IFT)

Paraffin oil was used in this study as an oil-phase. It is categorized as the light-oil that compatible with surfactants performance in the CEOR process which is effectively used in low to medium viscosity of oil. In addition, the aqueous phase consists of the distilled water and various concentrations of saponins were used to observe the performance of saponins at ambient conditions. Figure 3 shows the effect of saponin in surfactant solution to the interfacial tension.

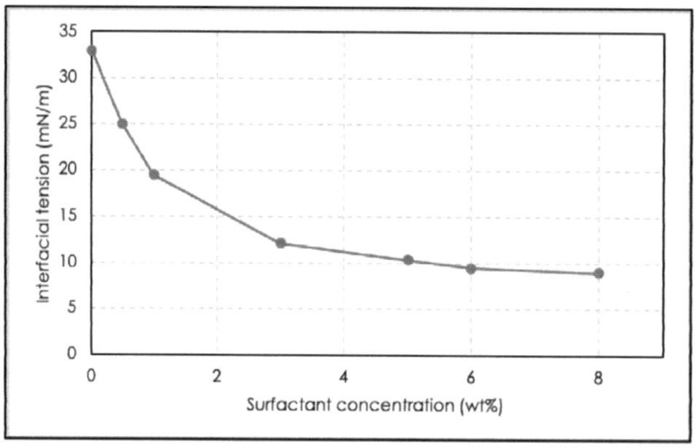

Figure 3. The effect of the saponin in surfactant solution to the interfacial tension.

This study proves that the IFT measurement of the saponins in surfactant flooding against paraffin oil showed a significant reduction. Figure 3 shows that the initial IFT between surfactant solution and paraffin oil is 33 mN/m which falls most significantly when reaching 0.5 wt% of saponin to become 25 mN/m and reaches a decrease of more than half after adding 3 wt% of saponin. The IFT reduction that occurred gradually reached 9 mN/m with an increase in saponin concentration of 8 wt%. This phenomena of IFT reduction occurs because the addition of surfactant concentration. This condition causes a single layer of surfactant monomer accumulation to interact more strongly interwoven by the surfactant structure with water-oil which is hydrophilic tail with paraffin oil and hydrophilic head with aqueous phase.

If the addition of surfactant concentration continues further, the interaction between water-surfactant-oil is getting stronger and the surfactant molecule gradually becomes denser towards the polar substance which resulting in the interfacial tension between water and oil to be more reduced. When the surfactant concentration reaches CMC, it causes the monomers to begin joining and forming the micelles. As concentration increases after CMC, micelle concentration increases without a significant increase in monomers concentration. As referred to this study, because of the water acts as a solvent, the concentration of surfactant above CMC is highly capable to dissolve oil rather than using water alone [22]. Figure 4 shows the comparison study on the effects of saponin addition towards interfacial tension using different saponin extraction sources.

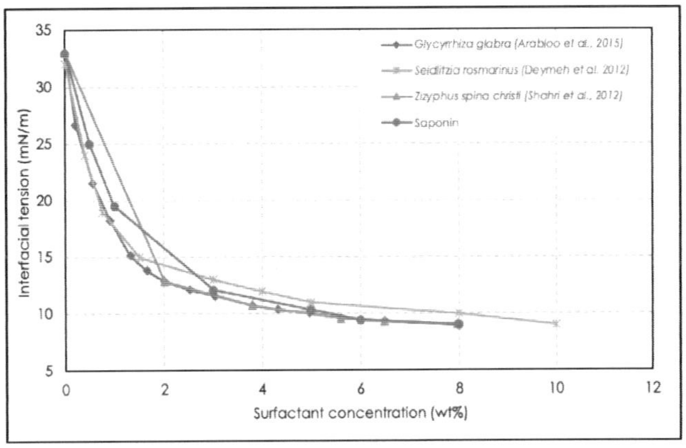

Figure 4. Comparison study on the effects of saponin addition towards interfacial tension using different saponin extraction sources [14,25,26].

The saponins addition in this study showed the same results with extracted saponins from *Seidlitzia rosmarinus*, *Glycyrrhiza glabra*, and *Zizyphus spina christi* under the same oil phase. Saponins with concentrations of 8 wt% of saponins, *Zizyphus spina christi*, and *Glycyrrhiza glabra* showed non-significant difference with the same results on IFT between surfactant flooding and paraffin oil which is 9 mN/m. Meanwhile, extracted saponins from *Seidlitzia rosmarinus* at a concentration of 8 wt% showed a significant difference compared to others saponin extraction sources with an IFT of 10 mN/m and succeeded in achieving an IFT of 9 mN/m at a concentration of 10 wt%. As a comparison, the IFT reduction process in each saponin shows a different trend of graph. Based on figure 4, it can be observed that saponins from different sources which are *Seidlitzia rosmarinus*, *Glycyrrhiza glabra*, *Zizyphus spina christi* and saponin show a significant decrease in IFT which is also different in respective concentration of 0.75 wt%, 1.5 wt%, 2 wt%, and 1 wt%. This phenomenon occurs due to the differences in the saponin structure of each plant-source. The interfacial thickness in surfactant flooding and oil will change in response to structural differences in surfactants which are closely related to the micelle formation process during CMC which render the differences in IFT value.

3.3. Wettability alteration

The natural wetting state in the sandstone which is water wet is able to change through the drainage process that involves the aging process using oil. This stage was performed to ensure the ability of saponins to change oil-wet state conditions into water-wet or tends to be more water wetness. Therefore, oil-wet sandstone performed under imbibition process by immersing the surfactant solution with saponin content in accordance with its CMC value. The concentration of saponins in CMC is intended to be the most efficient concentration both in terms of economy and the effect on oil mobilization. Figure 5 shows the effect of saponin in surfactant solution on contact angle of oil-wet sandstone.

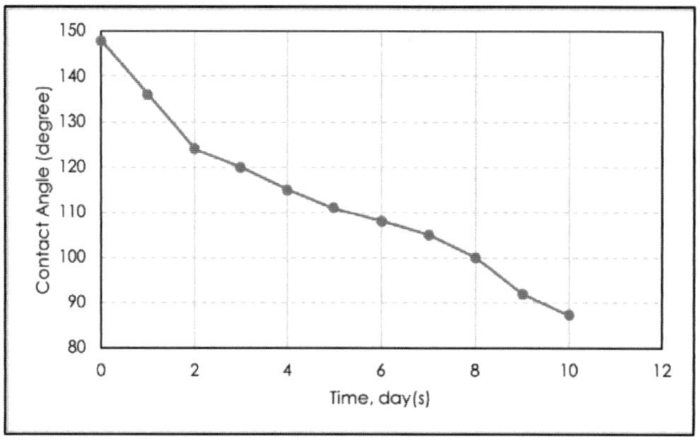

Figure 5. The effect of saponin in surfactant solution on contact angle of oil-wet sandstone.

Based on figure 5, the contact angle for distilled water only is 148°. Decrement in contact angle observed in about 16.2% on the second day with a value of 124°. On the following day, the decrease in contact angle is not more than 10° per day until the tenth day when the contact angle reaches less than 90° which is 87° as indicated to be an intermediate-wet state since the final contact angle is still in the range of 60°-120°. The proposed reason of the wettability alteration of the oil-wet sandstone prevail through the adsorption process caused by hydrogen bonding and/or hydrophobic interaction between surfactants and adsorbed paraffin oil on sandstone rock which not provided any evidence in this study [23,24].

4. Conclusion

Saponin has relatively low CMC value which is favorable to become a reference concentration in CEOR technique as natural surfactant. Based on the experimental results on surface tension at different saponin concentrations, the minimum surface tension and breakpoint of two slope values achieved the concentration of 0.05 wt% which is indicated as CMC of saponin in this study.

Saponin as natural surfactant is effective to alter the reservoir rock and fluid properties, facilitate oil mobilization, increase oil recovery and suggested as an additive to mitigating wax deposition. A 27% reduction from initial IFT between water and paraffin oil achieved by 8 wt% of saponin concentration. Observation from the IFT reduction also proved the effectiveness of saponins that can change oil-wet sandstone to intermediate-wet sandstone by the additional of saponins at CMC values.

Acknowledgement

The authors would like to express their sincere gratitude to Universiti Teknologi Malaysia Transdisciplinary Research Grant (Q.J130000.3551.07G12) for financial support in order for us to complete this research.

References

[1] Chhetri A B, Watts K C, Rahman M S and Islam M R 2009 Soapnut extract as a natural surfactant for enhanced oil recovery *Energy Sources, Part A Recover. Util. Environ. Eff.* **31** 1893–903

[2] Lashari N and Ganat T 2020 Emerging applications of nanomaterials in chemical enhanced oil recovery: Progress and perspective *Chinese J. Chem. Eng.* **28** 1995–2009

[3] Salehi M, Johnson S J and Liang J T 2008 Mechanistic study of wettability alteration using surfactants with applications in naturally fractured reservoirs *Langmuir* **24** 14099–107
[4] Kumar S and Mahto V 2017 Emulsification of Indian heavy crude oil using a novel surfactant for pipeline transportation *Pet. Sci.* **14** 372–82
[5] El-Sheshtawy H S and Khidr T T 2016 Some biosurfactants used as pour point depressant for waxy egyptian crude oil *Pet. Sci. Technol.* **34** 1475–82
[6] Wang Z, Yu X, Li J, Wang J and Zhang L 2016 The use of biobased surfactant obtained by enzymatic syntheses forwax deposition inhibition and drag reduction in crude oil pipelines *Catalysts* **6**
[7] Okoliegbe IN and Agarry OO 2012 Application of microbial surfactant (a review) *Sch. Journals Biotechnol.* **1** 15–23
[8] Akinyemi O P, Udonne J D, Efeovbokhan V E and Ayoola A A 2016 A study on the use of plant seed oils, triethanolamine and xylene as flow improvers of Nigerian waxy crude oil *J. Appl. Res. Technol.* **14** 195–205
[9] Chen G, Bai Y, Zhang J, Yuan W, Song H and Jeje A 2016 Synthesis of new flow improvers from canola oil and application to waxy crude oil *Pet. Sci. Technol.* **34** 1285–90
[10] De S, Malik S, Ghosh A, Saha R and Saha B 2015 A review on natural surfactants *RSC Adv.* **5** 65757–67
[11] Zhou W, Yang J, Lou L and Zhu L 2011 Solubilization properties of polycyclic aromatic hydrocarbons by saponin, a plant-derived biosurfactant *Environ. Pollut.* **159** 1198–204
[12] Stanimirova R, Marinova K, Tcholakova S, Denkov N D, Stoyanov S and Pelan E 2012 Surface shear rheology of saponin adsorption layers *Langmuir* **28** 12071–84
[13] Chen W J, Hsiao L C and Chen K K Y 2008 Metal desorption from copper(II)/nickel(II)-spiked kaolin as a soil component using plant-derived saponin biosurfactant *Process Biochem.* **43** 488–98
[14] Shahri M P, Shadizadeh S R and Jamialahmadi M 2012 A new type of surfactant for enhanced oil recovery *Pet. Sci. Technol.* **30** 585–93
[15] Moradi S, Isari A A, Bachari Z and Mahmoodi H 2019 Combination of a new natural surfactant and smart water injection for enhanced oil recovery in carbonate rock: Synergic impacts of active ions and natural surfactant concentration *J. Pet. Sci. Eng.* **176** 1–10
[16] Pal N, Samanta K and Mandal A 2019 A novel family of non-ionic gemini surfactants derived from sunflower oil: Synthesis, characterization and physicochemical evaluation *J. Mol. Liq.* **275** 638–53
[17] Alnoush W, Sayed A and Alyafei N 2019 Optimization of contact angle and interfacial tension measurements for fluid/rock systems at ambient conditions *MethodsX* **6** 1706–15
[18] Williams D L, Kuhn A T, Amann M A, Hausinger M B, Konarik M M and Nesselrode E I 2010 Computerised measurement of contact angles *Galvanotechnik* **101** 2502–12
[19] Hou B, Jia R, Fu M, Wang Y, Bai Y and Huang Y 2018 Wettability Alteration of an Oil-Wet Sandstone Surface by Synergistic Adsorption/Desorption of Cationic/Nonionic Surfactant Mixtures *Energy and Fuels* **32** 12462–8
[20] Garcia M T, Kaczerewska O, Ribosa I, Brycki B, Materna P and Drgas M 2017 Hydrophilicity and flexibility of the spacer as critical parameters on the aggregation behavior of long alkyl chain cationic gemini surfactants in aqueous solution *J. Mol. Liq.* **230** 453–60
[21] Mitra S and Dungan S R 1997 Micellar Properties of Quillaja Saponin. 1. Effects of Temperature, Salt, and pH on Solution Properties *J. Agric. Food Chem.* **45** 1587–95
[22] Kamal M S, Hussein I A and Sultan A S 2017 Review on Surfactant Flooding: Phase Behavior, Retention, IFT, and Field Applications *Energy and Fuels* **31** 7701–20
[23] Hou B F, Wang Y F and Huang Y 2015 Mechanistic study of wettability alteration of oil-wet sandstone surface using different surfactants *Appl. Surf. Sci.* **330** 56–64
[24] Zeng X and Osseo-Asare K 2004 Partitioning behavior of silica in the Triton X-100/dextran/water aqueous biphasic system *J. Colloid Interface Sci.* **272** 298–307

[25] Arabloo M, Ghazanfari M H and Rashtchian D 2016 Wettability modification, interfacial tension and adsorption characteristics of a new surfactant: Implications for enhanced oil recovery *Fuel* **185** 199–210
[26] Deymeh H, Shadizadeh S R and Motafakkerfard R 2012 Experimental investigation of Seidlitzia rosmarinus effect on oil-water interfacial tension: Usable for chemical enhanced oil recovery *Sci. Iran.* **19** 1661–4

Effects of Hot-Press Temperature and Mix Proportions on Mechanical Behaviours of Hybrid Coconut Fibre Particleboard

D N Tawasil[1*], E Aminudin [1,3], H A S Lim[1,3], L P Chung[2], N M Z N Soh[4]
[1] School of Civil Engineering, Faculty of Civil Engineering, Universiti Teknologi Malaysia, 81310 UTM Skudai, Johor Bahru, Malaysia
[2] Faculty of Built Environment and Surveying, Universiti Teknologi Malaysia, 81310 UTM Skudai, Johor Bahru, Malaysia
[3] UTM Construction Research Centre, Universiti Teknologi Malaysia, 81310 UTM Skudai, Johor Bahru, Malaysia
[4] Department of Civil Engineering & Technology, Faculty of Civil & Environmental Engineering, Universiti Tun Hussein Onn Malaysia, 86400 Parit Raja, Johor, Malaysia
Email: wannatawasil@gmail.com

Abstract. This paper represents the process in selecting the suitable hot press temperature, resin content and percentage of proportion for sawdust and coconut fibre in producing particleboard by investigating the mechanical behaviours (BS EN 310: 1993) and physical properties (BS EN 323:1993, BS EN 324). The process begins with preparation of materials which are sawdust and coconut fibre that were sieved and retained at 5mm sieve with oven-dried. The mixing process of this hybrid particle board which consist from sawdust and coconut fibre with the additional of resin (Urea Formaldehyde) being sprayed and hot pressed. The thickness and density for this experiment were fixed with targeted at 16mm and 650 kg/m^3 respectively. The hot press temperature was manipulated varies from 140 °C, 160°C and 180°C, resin content varies from 6%, 8% and 10%, while the percentage of proportion (%) of sawdust (SD) to coconut fibre (CF) varies from 100SD:0CF, 70SD:30CF, 50SD:50CF, 30SD:70CF and 0SD:100CF of weight. The particleboards were conditioned to room temperature for 7 days before tested for physical and mechanical properties. The results show that the suitable hot press temperature is 160°C, resin content at 8% and percentage of proportion for sawdust to coconut fibre are 30SD:70CF and 0SD:100CF for particleboard where the Modulus of Elasticity (MOE) 2689 N/mm² and 2450 N/mm² respectively, Modulus of Rupture (MOR) are 18.74 N/mm² and 20.69 N/mm² respectively in accordance to BS EN 312:2010.

1. Introduction

Different sources of natural fibres, as alternatives to concrete formulation, are explored which can be divided into two categories: farm or agricultural waste and commodity crops. For example, agriculture waste types of natural fibres that have been used in the formulation of green concrete and other building materials are bagasse, bamboo leaf ash, coconut coir, rice husks, palm oil fuel ash, wood chips, wheat straw, bananas, stroke and sisal; while natural fibre from commodities crops such as kenaf, jute and hemp [1].

The fibre of coconut is widely available. Various investigations were carried out into the use of coconut fibre in a matrix polymer composite. It has a less thermal conductivity and a good weight to strength ratio. It is environmentally friendly. Coconut fibre is a brilliant alternative to environmentally harmful materials the use of the building material [2]. From external shell of coconut fruit, the coconut fibre also known as coir is extracted. Coconut is mainly grown in tropical and sub-tropical regions. Ghana produces approximately 305,000 tonnes of coconut annually, and it generates many waste throughout the country. Coconut fibres are usually available in three main ways, (1) long fibre bristle, (2) short fibre mattress, and (3) decorticated (mixed fibre lengths). The engineering application of coconut fibre is not well known, but it has good engineering properties. The dimensions of coconut fibre are different and are thought to depend on the type, location and maturity of the coconut plant. The length to diameter (aspect proportion) of the fibre, which determines its application, is affected the flexibility and breakup of the fibre. The main compositions of the coconut fibre are cellulose, hemicellulose and lignin, which affects both the physical and mechanical characteristics of the fibre [3].

Coconut fibre has been chosen in this study due to its material, which contain of hardest natural fibre, with the percentage of lignin content are more than 30 percent [4] as shown in Table 1. The function of lignin to make the fibres tougher and more rigid by supplying compressive strength to the tissue and fibre as well as tightening up the cell wall to protect the carbohydrates against chemical or physical harm. Coconut fibres can be 4-6 times higher in strain than other fibres. In addition, coconut fibre does not lose its strength on storage and expose to sunlight. The coconut fibre density is important in determining the overall weight necessary for the coconut fibreboard panel. Due to its properties, the use of coconut fibre in furniture manufacturing is growing. Research has shown that coconut fibre is stronger than wood fibre [4].

Table 1. Chemical Composition and Physical Properties of Coconut Fibre [4]

Chemical Composition (%)	Physical Properties
Halocellulose (56.3%)	Density = 1.2 g/cm^3
α-cellulose (44.2%)	Elongation at break = 30%
Lignin (32.8%)	Tensile Strength =175 MPa
Ash (2.2%)	Young Modulus = 4 to 6 GPa
	Water Absorption = 130 -180%

The production of fibreboards from coconut husks has been studied by A. Lucia et. al in 2004 and discovered that coconut coir and fibre have high percentage of lignin, which is more than 130 °C hot thermally durable and can produce fibreboards without any chemical binders [5]. Recently, particleboards have different properties, including maximum design flexibility for easy production lines, reliable quality, dimensional variety, physical characteristics that are easy to use. There are a lot of application of particleboard as furniture, insulation material such as sound and thermal insulation, wall panel, wall bracing, partition and other industrial products. The characteristics of particleboards are influenced by many factors, including wood species, fibre structures, density, hardness, compressibility, particles type and size, and particle drying technology [6].The medium-size particle demonstrated the best performance for binder-less and melamine urea formaldehyde (MUF) bonded boards. When 16% MUF were used the bonded board of MUF showed the best performance. According to Ahmed (2016), the quality of the product with 16% MUF greater than binder-less coir pith board. Coconut coir board can provide sustainable, affordable and durable materials for construction and packaging as well as wood substitutes [7]. Urea formaldehyde is also used for indoor panels at moderate costs. However, these binders are not waterproof and emit formaldehyde toxic and carcinogenic. All these binders present problems for health and the environment. A reduced use of formaldehyde induces bad mechanical features of particle boards is well known. As formaldehyde reduction does not solve the toxic emissions, this adverse effect is therefore offset by plating or chemical changes [8].

According to El-Kassas et. al (2013), the properties of the produced fibreboards are dependent on the average density and the resin contents [9]. In his study, Akinyemi (2016) observed that the panel density increased as sawdust composition declined, up to the peak at 25% and then a drastically drop in value. However, as the proportion of corn cob increased, the density of particleboard slightly increased but a significant drop was observed after peak at 75% corn cob. Compared to the sawdust panels, the corn cob panels were of lower density. The density obtained is comparable to densities of particleboards in wood production industries of 590 and 800 kg/m^3 [10]. While in her study, Paridah (2014) found out that particleboards with 10% resin and 50:50 (RW:KS) have the highest strength (19.08 MPa) while particleboards with 70:30 (RW:KS) show a higher strength (2.23 GPa). The RW:KS ratio has more impact than the resin level on the thickness swelling (TS) and water absorption (WA). Hybrid particleboards made of 70% RW and 30% KS, 10% resin content shows all good property and is equivalent to 100% RW (control) samples. It concluded that kenaf stem can substitute rubber-wood particles of up to 50%, but resin levels must be maintained at 10% and higher because a lower level of resin (68%) dramatically reduces particleboards' strength [11].

The pre-pressed mats were pressed for 5 min in a hot press and the temperature and pressure were controlled at 160 °C and 3 N/mm^2 [7]. The pressing cycle was then conducted using the daylight press at pressing plate temperature of 180 °C and under adequate pressure to achieve the targeted panel density and thickness. The panel was pressed for a time of 12–13 s/mm of the panel thickness [9]. The pressing process has the following time, temperature and pressing specifications: i) for the HDF fibreboard, 220 °C and 320 kgf/cm^2 for 4 minutes; ii) for the MDF fibreboard, 210 °C and 320 kgf/cm^2 for 4 minutes.; iii) and for the MDF UF fibreboard, 160°C and 100 kgf/cm^2 for 10 minutes.

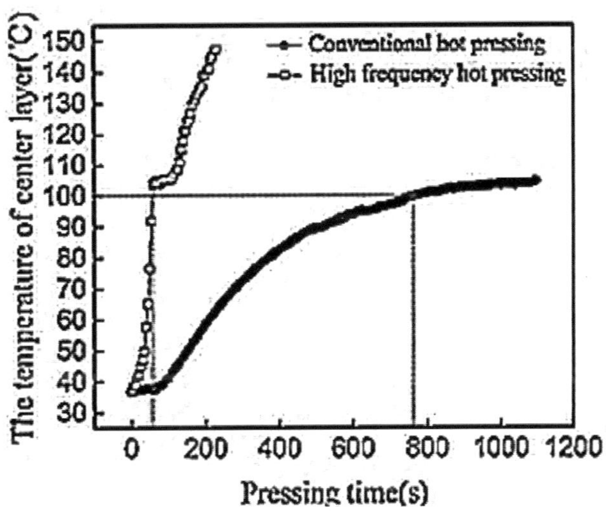

Figure 1. The graph of temperatures of center layer against pressing time [12]

2. Materials and Methodology

Sawdust particle and resin used in this study are courtesy from Heveaboard Sdn Bhd. The urea-formaldehyde (UF) resin with a solid content of 64 ± 1% was used for the experimental panels. This UF resin is currently used for the particleboard manufacturing on an industrial scale and has a dynamic viscosity of 150 to 250 cps at 30 °C and gel time reactivity of 120 to 160 s at 100 °C. Coconut fibre were prepared by shredding and crushing long fibres by using hammer mill crushing machine. After that, the coconut fibre sieve using horizontal handmade wood sieve of 5 mm, 3mm,1 mm and pan as well as for sawdust. After that, sawdust and coconut fibre retained at 5 mm will be separated for board making and the particle size analysis using vertical sieve shaker. Both sawdust and coconut fibres then were oven-dried at 80 °C for 24h to reduce the moisture content of materials.

The sawdust, coconut fibre and resin will be weighed to respective percentage proportion before to be mixing and blending together in drum mixer for 10 minutes. The design of percentage proportion of sawdust and coconut fibre are based of targeted density. The target density of the experimental panels was 650 kg/m³, which is the average density of the 16-mm particleboard manufactured in industrial conditions from wood particles. UF resin was added to the single-mat configuration at a level of 6 wt%, 8 wt% and 10 wt%, based on the weight of the particles.

Fifteen types of panel were made, as indicated in Table 2. The panels were prepared under laboratory conditions using a hot press with plates of 450 mm x 450 mm in Timber Fabrication Laboratory, Universiti Tun Hussein Onn Malaysia. The particleboards were manually formed in wooden frames whose dimensions are 350 mm x 350 mm x 16 mm. Boards measuring 350 mm x 350 mm were then hot-pressed at 140 °C, 160 °C and 180 °C for 7 minutes under 2.5 to 3.0 N/mm² pressure. After pressing, the particleboards were conditioned at 20 °C and 65% relative humidity for 1 week before evaluating the physical and mechanical properties. The particleboards were first trimmed to avoid edge defects to a final size of 300 mm x 300 x 16 mm. Three replicate panels were made for each board type, and specimens for mechanical and physical testing were cut from each particleboard type.

Table 2. Experimental Design for Board

Variable	**Board Code**	**% Proportion**		**Adhesive (UF) Resin (%)**			**Hot-press Temperature (°C)**		
		Sawdust	**Coconut fibre**	**6**	**8**	**10**	**140**	**160**	**180**
1	100SD (1)	100	0		√		√	√	√
(Different	70SD (1)	70	30		√		√	√	
Hot- Press	50SD (1)	50	50		√		√	√	
Tempera-	30SD (1)	30	70		√		√	√	
ture)	0SD (1)	0	100		√		√	√	√
	100SD (2)	100	0	√				√	
	70SD (2)	70	30	√				√	
	50SD (2)	50	50	√				√	
2	30SD (2)	30	70	√				√	
(Different	0SD (2)	0	100	√				√	
Resin	100SD (3)	100	0			√		√	
Content)	70SD (3)	70	30			√		√	
	50SD (3)	50	50			√		√	
	30SD (3)	30	70			√		√	
	0SD (3)	0	100			√		√	

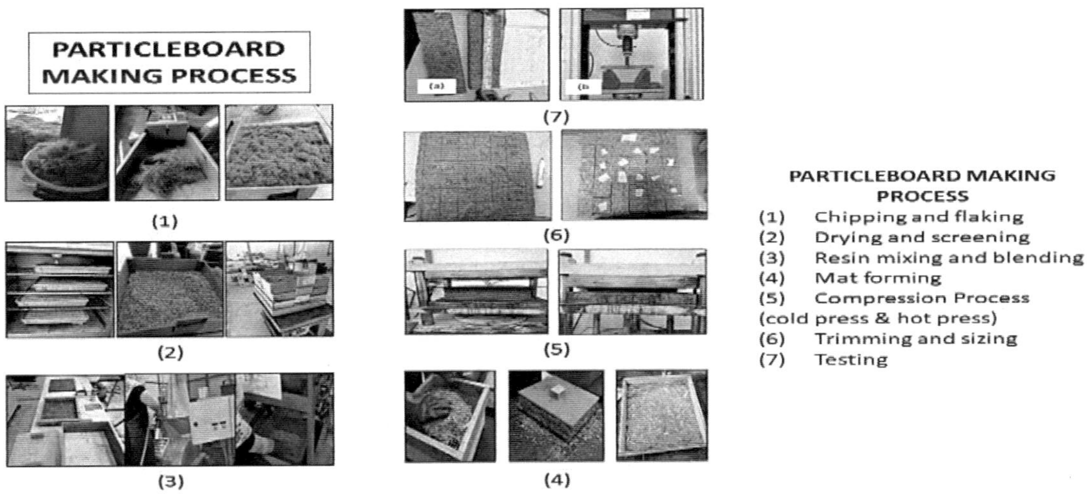

Figure 2. Hybrid Fibre Particleboard Making Process [13]

2.1. *Determination of Density (BS EN 323:1993)*

By adapting the methodology in BS EN 323:1993, the density to be determined by sampling and cutting of the test pieces in accordance to EN 326-1 where the side of nominal length of specimen is 50 mm. The dimensions and thickness at a point of the intersection of diagonals of each test piece should be measured in accordance to EN 325. After that, weighed the mass of each test piece. The density of a board shall be obtained by calculating the arithmetic mean of the densities of all the test pieces taken from the same board and is expressed in kg/m^3 to three significant figures [14]. The density ρ of each test piece (in kg/m^3) shall be calculated from the equation (1):

$$\rho = \frac{m}{b_1 \, x \, b_2 \, x \, t} x \; 10^6 \tag{1}$$

Where, m is the mass of board, in gram, b_1, b_2 is the dimension of board, in millimetre and t is the thickness of board, in millimetre.

2.2. *Determination of Particleboard's Modulus of Elasticity and Modulus of Rupture (BS EN 310:1993)*

In accordance to BS EN 310:1993, the modulus of elasticity, E_m, can be calculated by using equation (2) by sampling and cutting of the test pieces and measured the width and thickness of each test piece according to EN 325 at the following points: the thickness at the intersection of the diagonals; the width at the mid-length whereas the width and thickness of test pieces are 50 mm and as specified in thickness group (16 mm, 20 mm and 32 mm). Adjust the distance between the centres of the supports, to within 1 mm of 20 times the nominal thickness of the panel, but not less than 100 mm and not more than 1 000 mm, in this study the distance is fixed to 200 mm. The load shall be applied at a constant rate of cross-head movement throughout the test. The rate of loading shall be adjusted so that the maximum load is reached within (60 ± 30) s. Measure the deflection in the middle of the test piece (below the loading head) [15].

$$E_m = \frac{(l_1)^3 (F_2 - F_1)}{4 \, bt^3 (a_2 - a_1)} \tag{2}$$

Where, l_1 is the distance between the centres of the supports, in millimetres, b is the width of the test piece, in millimetres, t is the thickness of the test piece, in millimetres, $F_2 - F_1$ is the increment of

IOP Conf. Series: Materials Science and Engineering 1153 (2021) 012019 doi:10.1088/1757-899X/1153/1/012019

load on the straight line portion of the load-deflection curve, in N which F_1 shall be approximately 10 % and F_2 shall be approximately 40 % of the maximum load and $a_2 - a_1$ is the increment of deflection at the mid-length of the test piece (corresponding to $F_2 - F_1$).

The bending strength of each test piece shall be expressed to three significant figures. The bending strength for each group of test pieces taken from the same board is the arithmetic mean of the bending strengths of the appropriate test pieces, expressed to three significant figures [15]. The bending strength f_m (in N/mm^2), of each test piece, is calculated from the equation (3),

$$f_m = \frac{3\,F_{max}\,l_1}{2\,bt^2} \tag{3}$$

Where, F_{max} is the maximum load, in Newtons and l_1, b, and t are in millimetres.

3. Results and Discussions

The results obtained from this research are the density distribution, modulus of elasticity and modulus of rupture in all particleboards when the hot-press temperature varies from 140 °C, 160 °C and 180 °C as well as the results for modulus of elasticity and modulus of rupture in comparison between 100SD:0CF to 30SD:70CF and 0SD:100CF when the resin content varies from 6%, 8% and 10%. The results for modulus of rupture and modulus of elasticity were collected and presented in a bar graph as a comparison to requirements for non load-bearing boards for use in humid conditions (Type P3) particleboard from Table 4 in BS EN 312:2010.

3.1. Density of Board

Density distribution in particleboard are the mean from specimens as specified in Figure 3. After that the result represented in a bar graph as shown in Figure 3. First of all, the comparison between 100SD:0CF and 0SD:100CF when the hot-press temperature is 180 °C the density of particleboard are 513.526 kg/m^3 and 675.416 kg/m^3 respectively with differences of 136.474 kg/m^3 and 25.416 kg/m^3 where the particleboard with coconut fiber is more stable in density distribution. After that, when the hot-press temperature is 140 °C all particleboards have density more than targeted density of 650 kg/m^3 where 702.963 kg/m^3, 706.971 kg/m^3, 696.702 kg/m^3, 759.185 kg/m^3 and 669.705 kg/m^3 which differences range from 19.705 kg/m^3 to 109.185 kg/m^3. Meanwhile, when the hot-press temperature is 160 °C the density for all particleboards are 614.720 kg/m^3, 666.737 kg/m^3, 613.451 kg/m^3, 696.383 kg/m^3, 667.721 kg/m^3 which differences range from 16.737 kg/m^3 to 46.383 kg/m^3.

In can be concluded that the range of differences of the density of particleboards when the hot-press temperature is 160 °C suitable in particleboard making because the density distribution more stable. According to E. Dieckmann et al. (2019), The pressing temperature should be between 150 and 160 °C, is crucial for developing optimal properties. [16].

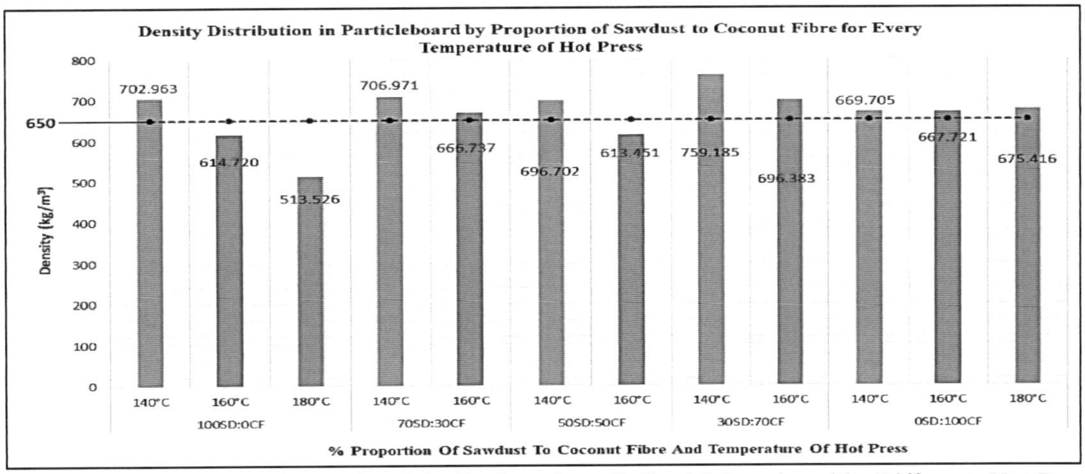

Figure 3. Bar Graph of Density Distribution for all Particleboards with Different Hot-Press Temperature.

3.2. Modulus of Elasticity in Bending and of Bending Strength (Modulus of Rupture) of Boards

Modulus of elasticity (MOE) in bending for all particleboards were tested and then compared in Figure 4. First of all, the comparison between 100SD:0CF and 0SD:100CF with hot press-temperature of 160 °C, both particleboards reached 2677 N/mm^2 and 2450 N/mm^2 respectively which passed 1950 N/mm^2, however with hot-press temperature of 180 °C the particleboard 0SD:100CF reached 2197.500 N/mm^2 which also passes the requirement for Type P3 particleboard while 100SD:0CF only reached up to 695.450 N/mm^2. Therefore, with hot-press temperature of 160 °C and the more percentage of coconut fiber in particleboard the better the MOE in bending. After that, the comparison between 70SD:30CF, 50SD:50CF and 30SD:70CF with hot-press temperature of 160 °C, only particleboard 70SD:30CF and 30SD:70CF passed the requirement which are reached 2740.500 N/mm^2 and 2689.000 N/mm^2 respectively while 50SD:50CF only reached up to 931.300 N/mm^2. Finally, when the hot-press temperature is 140 °C all particleboard have not passed the requirement where the MOE for 100SD:0CF, 70SD:30CF, 500SD:50CF, 30SD:70CF and 0SD:100CF are 1568.333 N/mm^2, 1576.833 N/mm^2, 963.667 N/mm^2, 1741.500 N/mm^2 and 893.833 N/mm^2 respectively.

Therefore, when the percentage of replacement of sawdust to coconut fibre as well as replacement of coconut fibre to sawdust reach 30% and the hot-press temperature of 160 °C the boards achieved and passes 1950 N/mm^2. Since the failure process varies depending on the pressing temperature, this is the case where peak fibre bonding induces fibre pull out at low temperatures (around 140 °C). Under perfect pressing conditions, fibre fracture and heavy bonding occur. Due to thermal erosion of feather fibres, brittle fracture happens at processing temperatures above 170 °C. [16]. According to Gul et al. (2017), the temperature range of 160–180°CC is typically used to ensure complete solidification of different glues [17].

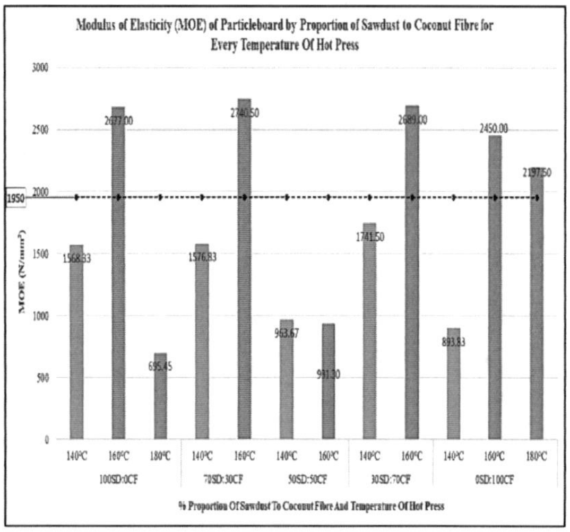

Figure 4. Bar Graph of Modulus of Elasticity for all Particleboards with Different Hot-Prcss Temperature

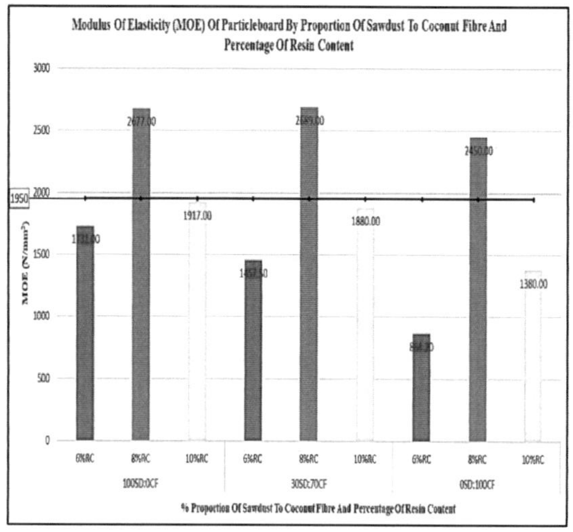

Figure 5. Bar Graph of Modulus of Elasticity for all Particleboards with Different Resin Content

Figure 5 shows the comparison between particleboards 100SD:0CF with 30SD:70CF and 0SD:100CF with different resin content varies from 6%, 8% and 10%. With 6% of resin content all particleboards do not passes the requirement where the MOE only reached 1731.00 N/mm^2, 1457.50 N/mm^2 and 864.20 N/mm^2 for 100SD:0CF, 30SD:70CF and 0SD:100CF respectively. Meanwhile, at 8% of resin content all the particleboards passed the requirement by reached over 1950 N/mm^2 whereas 2677.00 N/mm^2 for 100SD:0CF, 2689.00 N/mm^2 for 30SD:70CF and 2450.00 N/mm^2 for 0SD:100CF. At last but not least, at 10% of resin content all particleboard does not passes the requirement by only reached 1917.00 N/mm^2 for 100SD:0CF, 1880.00 N/mm^2 for 30SD:70CF and 1380.00 N/mm^2 for 0SD:100CF. Therefore, 8% of resin content is suitable for particleboards to reached 1950 N/mm^2 as well as the 30SD:70CF reach the highest value of all is the suitable ratio proportion of sawdust and coconut fibers as a particleboard.

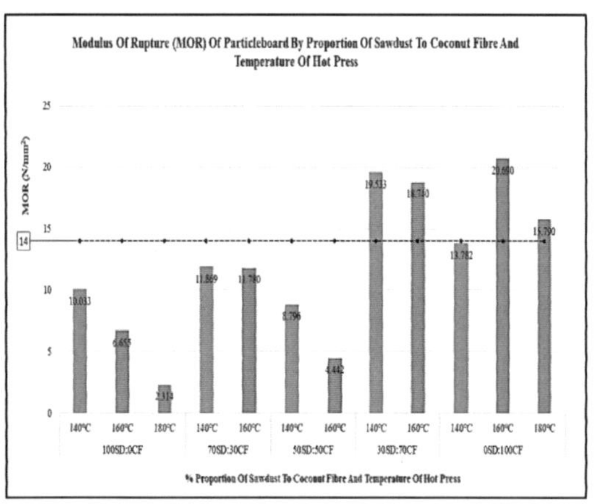

Figure 6. Bar Graph of Modulus of Rupture for all Particleboards with Different Hot-Press Temperature

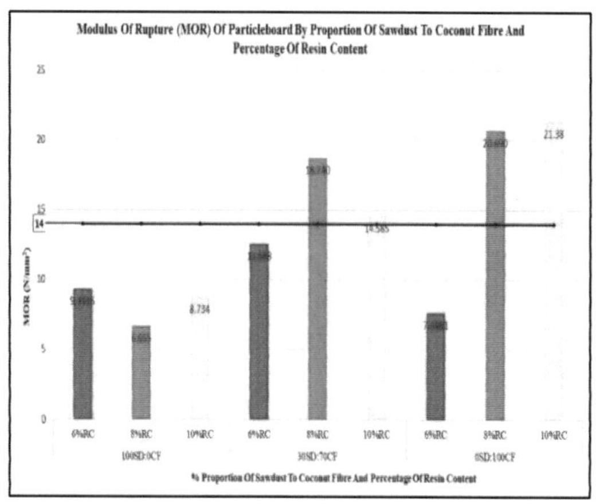

Figure 7. Bar Graph of Modulus of Rupture for all Particleboards with Different Resin Content

Bending strength or Modulus of rupture for all particleboards were tested and then compared in Figure 6. First of all, the comparison between 100SD:0CF and 0SD:100CF with hot press-temperature of 160 °C, only particleboard 0SD:100CF reached 20.690 N/mm^2 which passed 14 N/mm^2 while 100SD:0CF only reached up to 6.655 N/mm^2, however with hot-press temperature of 180 °C the particleboard 0SD:100CF reached 15.790 N/mm^2 which also passes the requirement for Type P3 particleboard while 100SD:0CF only reached up to 2.314 N/mm^2. Therefore, with hot-press temperature of 160 °C and the more percentage of coconut fiber in particleboard the better the modulus of rupture. After that, the comparison between 70SD:30CF, 50SD:50CF and 30SD:70CF with hot-press temperature of 160 °C, only particleboard 30SD:70CF passed the requirement which are reached 18.740 N/mm^2 while 70SD:30CF and 50SD:50CF only reached up to 11.780 N/mm^2 and 4.442 N/mm^2 respectively. Finally, when the hot-press temperature is 140 °C only particleboard 30SD:70CF passes the requirement which reached 19.533 N/mm^2 while other have not passed the requirement where the MOE for 100SD:0CF, 70SD:30CF, 500SD:50CF and 0SD:100CF are 10.033 N/mm^2, 11.869 N/mm^2, 8.796 N/mm^{2}and 13.782 N/mm^2 respectively. According to Gul et. Al (2017), even at low temperature (140 °C), the MOR value is very close to standard value. At 160 °C, the MOR value is at its peak and if the temperature further to increase from 160 °C to 185 °C, the MOR behaves in reverse manner [17].

Figure 7 shows the comparison of bending strength or modulus of rupture between particleboards 100SD:0CF with 30SD:70CF and 0SD:100CF with different resin content varies from 6%, 8% and 10%. With 6% of resin content all particleboards do not passes the requirement where the MOR only reached 9.3115 N/mm^2, 12.588 N/mm^2 and 7.6481 N/mm^2 for 100SD:0CF, 30SD:70CF and 0SD:100CF respectively. Meanwhile, at 8% of resin content only two particleboards passed the requirement by reached over 14 N/mm^2 whereas 18.740 N/mm^2 for 30SD:70CF and 20.690 N/mm^2 for 0SD:100CF while 6.655 N/mm^2 for 100SD:0CF do not passes the requirement. At last but not least, at 10% of resin content only particleboard 100SD:0CF does not passes the requirement by reached 8.734 N/mm^2, meanwhile both 30SD:70CF and 0SD:100CF passed with 14.585 N/mm^2 and 21.38 N/mm^2 respectively. Therefore, 8% and 10% of resin content is suitable for particleboards to reached 14 N/mm^2 as well as particleboard 0SD:100CF reached the highest value of all is the suitable ratio proportion of sawdust and coconut fibers as a particleboard as well as 30SD:70CF as the more the proportion of coconut fibre the better the bending strength obtained.

4. Conclusion

All the particleboards of sawdust and coconut fibre were tested and compared with BS EN 312:2010 requirement for non load-bearing boards for use in humid condition (Type P3). The hot-press temperature used during compression varied from 140 °C, 160 °C and 180 °C, while the resin content varied from 6%, 8% and 10% and the ratio of proportion sawdust to coconut fibre by percentage of weight varied from 100SD:0CF, 70SD:30CF, 505SD:50CF, 30SD:70CF and 0SD:100CF. It is concluded that the suitable hot-press temperature is 160 °C, resin content is 8% and the proportion of sawdust to coconut fibre are 30SD:70CF and 0SD:100CF whereas the more the fibre the better the results obtained.

Acknowledgement

The authors would like to thank Universiti Teknologi Malaysia for supporting with financial grants, Cost Centre No: Q.J130000.3551.06G64, Q. J130000.3552.07G55, and GreenPROMPT research team members and Department of Construction Management, UTM CRC, together with opportunity to conduct the research.

References

[1] M. A. Alam, S. M. Sapuan, and M. R. Mansor, *20 - Design characteristics, codes and standards of natural fibre composites*. Elsevier Ltd, 2017.

[2] P. M. K, M. K. S, R. K. V, and H. K. Govindaraju, "ScienceDirect Characterization of

Mechanical Properties of Epoxy Reinforced with Glass Fiber and Coconut Fiber," *Mater. Today Proc.*, vol. **16**, pp. 661–667, 2019, doi: 10.1016/j.matpr.2019.05.143.

[3] H. Danso, "ScienceDirect ScienceDirect Properties of Coconut , Oil Palm and Bagasse Fibres : As Potential Building Materials," *Procedia Eng.*, vol. **200**, pp. 1–9, 2017, doi: 10.1016/j.proeng.2017.07.002.

[4] N. Ngadiman, M. Kaamin, A. A. Kadir, S. Sahat, and A. Zaini, "Panel Board From Coconut Fibre And Pet Bottle," vol. **01014**, pp. 1–9, 2018.

[5] A. Lúcia *et al.*, "Environmental assessment of bioproducts in development stage: The case of fiberboards made from coconut residues," *J. Clean. Prod.*, 2017, doi: 10.1016/j.jclepro.2017.03.100.

[6] A. A. Adediran *et al.*, "Properties of agro-based hybrid particleboards," *Procedia Manuf.*, vol. **35**, pp. 442–446, 2019, doi: 10.1016/j.promfg.2019.05.064.

[7] W. T. Discipline and K. Luang, "Particleboard from coir pith," vol. 51, no. 3, pp. 239–245, 2016.

[8] V. Pontevedra and J. Moores, "ScienceDirect ScienceDirect ScienceDirect Particleboard of Iroko and African Locust Bean Pod capacity optimization in DL Development and Investigation into Properties of composite between used capacity and operational efficiency Liverpool much," *Procedia Manuf.*, vol. **30**, pp. 188–193, 2019, doi: 10.1016/j.promfg.2019.02.027.

[9] A. I. Mourad, "Novel fibers preparation technique for manufacturing of rice straw based fiberboards and their characterization," *J. Mater.*, 2013, doi: 10.1016/j.matdes.2013.03.057.

[10] A. B. Akinyemi, J. O. Afolayan, and E. O. Oluwatobi, "Some properties of composite corn cob and sawdust particle boards," *Constr. Build. Mater.*, vol. **127**, pp. 436–441, 2016, doi: 10.1016/j.conbuildmat.2016.10.040.

[11] M. T. Paridah, A. H. Juliana, Y. A. El-shekeil, M. Jawaid, and O. Y. Alothman, "Measurement of mechanical and physical properties of particleboard by hybridization of kenaf with rubberwood particles," *MEASUREMENT*, vol. **56**, pp. 70–80, 2014, doi: 10.1016/j.measurement.2014.06.019.

[12] K. Wei, C. Lv, M. Chen, X. Zhou, Z. Dai, and D. Shen, "Development and performance evaluation of a new thermal insulation material from rice straw using high frequency hot-pressing," *Energy Build.*, vol. **87**, pp. 116–122, 2015, doi: 10.1016/j.enbuild.2014.11.026.

[13] M. F. N. Dos Santos, R. A. G. Battistelle, B. S. Bezerra, and H. S. A. Varum, "Comparative study of the life cycle assessment of particleboards made of residues from sugarcane bagasse (Saccharum spp.) and pine wood shavings (Pinus elliottii)," *J. Clean. Prod.*, vol. **64**, no. March, pp. 345–355, 2014, doi: 10.1016/j.jclepro.2013.06.039.

[14] British Standard, "Wood-based panels - Determination of density," pp. 1–12, 1993.

[15] CEN, "EN 310 Wood-based panels - Determination of modulus of elasticity in bending and of bending strength," 1996.

[16] E. Dieckmann, K. Eleftheriou, T. Audic, K. Y. Lee, L. Sheldrick, and C. Cheeseman, "New sustainable materials from waste feathers: Properties of hot-pressed feather/cotton/bi-component fibre boards," *Sustain. Mater. Technol.*, vol. **20**, p. e00107, 2019, doi: 10.1016/j.susmat.2019.e00107.

[17] W. Gul, A. Khan, and A. Shakoor, "Impact of hot pressing temperature on Medium Density Fiberboard (MDF) performance," *Adv. Mater. Sci. Eng.*, vol. **2017**, no. January, 2017, doi: 10.1155/2017/4056360.

AUTHOR INDEX